Atmospheric Mercury Monitoring, Analysis, and Chemistry

Atmospheric Mercury Monitoring, Analysis, and Chemistry

New Insights and Progress toward Minamata Convention Goals

Editor

James Cizdziel

MDPI • Basel • Beijing • Wuhan • Barcelona • Belgrade • Manchester • Tokyo • Cluj • Tianjin

Editor
James Cizdziel
University of Mississippi
USA

Editorial Office
MDPI
St. Alban-Anlage 66
4052 Basel, Switzerland

This is a reprint of articles from the Special Issue published online in the open access journal *Atmosphere* (ISSN 2073-4433) (available at: https://www.mdpi.com/journal/atmosphere/special_issues/Atmospheric_Mercury_Monitoring).

For citation purposes, cite each article independently as indicated on the article page online and as indicated below:

LastName, A.A.; LastName, B.B.; LastName, C.C. Article Title. *Journal Name* **Year**, *Volume Number*, Page Range.

ISBN 978-3-0365-0774-3 (Hbk)
ISBN 978-3-0365-0775-0 (PDF)

Cover image courtesy of Winston Luke.

© 2021 by the authors. Articles in this book are Open Access and distributed under the Creative Commons Attribution (CC BY) license, which allows users to download, copy and build upon published articles, as long as the author and publisher are properly credited, which ensures maximum dissemination and a wider impact of our publications.

The book as a whole is distributed by MDPI under the terms and conditions of the Creative Commons license CC BY-NC-ND.

Contents

About the Editor vii

James V. Cizdziel
Atmospheric Mercury Monitoring, Analysis, and Chemistry: New Insights and Progress toward Minamata Convention Goals
Reprinted from: *Atmosphere* **2021**, *12*, 166, doi:10.3390/atmos12020166 1

Mae Sexauer Gustin, Sarrah M. Dunham-Cheatham, Jiaoyan Huang, Steve Lindberg and Seth N. Lyman
Development of an Understanding of Reactive Mercury in Ambient Air: A Review
Reprinted from: *Atmosphere* **2021**, *12*, 73, doi:10.3390/atmos12010073 5

Igor Živković, Sabina Berisha, Jože Kotnik, Marta Jagodic and Milena Horvat
Traceable Determination of Atmospheric Mercury Using Iodinated Activated Carbon Traps
Reprinted from: *Atmosphere* **2020**, *11*, 780, doi:10.3390/atmos11080780 23

Byunggwon Jeon, James V. Cizdziel, J. Stephen Brewer, Winston T. Luke, Mark D. Cohen, Xinrong Ren and Paul Kelley
Gaseous Elemental Mercury Concentrations along the Northern Gulf of Mexico Using Passive Air Sampling, with a Comparison to Active Sampling
Reprinted from: *Atmosphere* **2020**, *11*, 1034, doi:10.3390/atmos11101034 37

Marc W. Beutel, Lanka DeSilva and Louis Amegbletor
Direct Measurement of Mercury Deposition at Rural and Suburban Sites in Washington State, USA
Reprinted from: *Atmosphere* **2021**, *12*, 35, doi:10.3390/atmos12010035 55

Trajče Stafilov, Lambe Barandovski, Robert Šajn and Katerina Bačeva Andonovska
Atmospheric Mercury Deposition in Macedonia from 2002 to 2015 Determined Using the Moss Biomonitoring Technique
Reprinted from: *Atmosphere* **2020**, *11*, 1379, doi:10.3390/atmos11121379 67

Yi Tang, Qingru Wu, Wei Gao, Shuxiao Wang, Zhijian Li, Kaiyun Liu and Deming Han
Impacts of Anthropogenic Emissions and Meteorological Variation on Hg Wet Deposition in Chongming, China
Reprinted from: *Atmosphere* **2020**, *11*, 1301, doi:10.3390/atmos11121301 77

Samantha T. Brown, Lloyd L. Bandoo, Shenelle S. Agard, Shemeiza T. Thom, Tamara E. Gilhuys, Gautham K. Mudireddy, Arnith V. Eechampati, Kazi M. Hasan, Danielle C. Loving, Caryn S. Seney and Adam M. Kiefer
A Collaborative Training Program to Assess Mercury Pollution from Gold Shops in Guyana's Artisanal and Small-Scale Gold Mining Sector
Reprinted from: *Atmosphere* **2020**, *11*, 719, doi:10.3390/atmos11070719 89

Bruce Gavin Marshall, Arlette Andrea Camacho, Gabriel Jimenez and Marcello Mariz Veiga
Mercury Challenges in Mexico: Regulatory, Trade and Environmental Impacts
Reprinted from: *Atmosphere* **2021**, *12*, 57, doi:10.3390/atmos12010057 111

About the Editor

James Cizdziel is Associate Professor in the Department of Chemistry and Biochemistry at the University of Mississippi. His research interests are in the areas of analytical, environmental, and forensic chemistry. He has pioneered methods in trace elemental analyses, trace evidence from 3D-printed guns, atmospheric sampling using unmanned aerial vehicles, characterization of microplastics in environmental samples, and the biogeochemical cycling of mercury. He has authored more than 70 papers and has been funded by the US Environmental Protection Agency, Department of Energy, Department of Justice, and the National Science Foundation. He teaches courses in analytical chemistry, instrumental analysis, and applied spectroscopy, and enjoys working with students to develop new measurement techniques or applying existing methods in novel ways. When not in the laboratory, he enjoys fishing, golfing, tennis, and traveling.

Editorial

Atmospheric Mercury Monitoring, Analysis, and Chemistry: New Insights and Progress toward Minamata Convention Goals

James V. Cizdziel

Department of Chemistry and Biochemistry, University of Mississippi, University, MS 38677, USA; cizdziel@olemiss.edu

Citation: Cizdziel, J.V Atmospheric Mercury Monitoring, Analysis, and Chemistry: New Insights and Progress toward Minamata Convention Goals. *Atmosphere* 2021, 12, 166. https://doi.org/10.3390/atmos12020166

Received: 11 January 2021
Accepted: 26 January 2021
Published: 28 January 2021

Publisher's Note: MDPI stays neutral with regard to jurisdictional claims in published maps and institutional affiliations.

Copyright: © 2021 by the author. Licensee MDPI, Basel, Switzerland. This article is an open access article distributed under the terms and conditions of the Creative Commons Attribution (CC BY) license (https://creativecommons.org/licenses/by/4.0/).

Mercury is a persistent and toxic global contaminant that is transported through the atmosphere, deposits to terrestrial and aquatic ecosystems, and concentrates up the food chain reaching levels that can harm both humans and wildlife [1]. In this *Special Issue* on atmospheric mercury (Hg), seven original research articles and a review paper report the latest findings describing the distribution, deposition, and measurement of this airborne pollutant as well as the human and environmental impacts of artisanal mining of Hg and gold. The papers span a wide range of investigations including the determination of Hg deposition in northwest USA [2], southeast Europe [3], and Chongming Island, near Shanghai, China [4]; the development and use of iodinated activated carbon for traceable determination of airborne Hg [5]; new measurements of atmospheric Hg along the northern Gulf of Mexico using passive air sampling [6]; the impacts of artisanal gold mining and associated Hg trade in south and central America [7,8]; and a timely review on reactive Hg in ambient air [9]. This editorial provides highlights of these interesting papers and presents them in the broader context of modern atmospheric Hg research and the Minamata Convention on Mercury, a global treaty now signed by 127 parties and designed to protect human health and the environment from anthropogenic emissions of Hg.

Advances in measuring atmospheric Hg have improved our understanding of its sources, transport, transformations, fluxes, and fate. Airborne Hg is generally categorized as gaseous elemental mercury (GEM) and reactive mercury (RM), with the latter representing gaseous oxidized mercury (GOM) and particle-bound mercury (PBM). Accurately measuring RM as a whole is challenging, let alone individual Hg compounds. Gustin et al. [9] provide a review focusing on the fascinating history of efforts to quantify and characterize RM, along with the current state of such measurements and knowledge. Methods for measuring RM are changing as concerns over artifacts or repeatability emerge or as knowledge of RM improves. Early work using mist chambers [10], filters and membranes [11], and denuders [12] naturally resulted in inter-comparison studies, especially after an automated system became commercially available from Tekran® Instruments Corporation that improved temporal resolution. The Tekran speciation system was an important step forward because as Hg sources and oxidant chemistry of the air vary, so too do levels of RM. The Tekran, as it is known, became standard, and the system was employed by researchers worldwide. More recently, it was demonstrated that the KCl denuder used in the Tekran is not adequate for measuring RM concentration [9], and so the development of new methods continues. Among them are the Reactive Mercury Active System (RMAS) and the ever-promising mass spectrometry. Moving forward, the review authors strongly advocate for the field calibration of oxidized Hg measurements. The review concludes with useful sections on "what we have learned" and "work needed."

Continuing on the important theme of attaining accurate and reliable measurements, Zivkovic and colleagues [5] investigated spiking of a standard reference material (NIST 3133) directly onto iodinated activated carbon (AC) traps for accurate and traceable calibration determination of ambient Hg^0. Measurements were made by atomic absorption spectrometry after sample combustion. There were identical responses between Hg loaded directly and Hg loaded by purging from the NIST solution after reduction with $SnCl_2$ across a wide concentration range (10–2000 ng). Further proof-of-concept studies on real atmospheric samples, where accuracy was assessed by a different reference material, showed that the approach is, indeed, effective for measuring atmospheric Hg. The authors include details for preparation and optimization of the quantitative analysis setup. They also deconvoluted peaks of fractionation thermograms to identify iodide of Millon's base ($HgO·Hg(NH_3)_2I_2$) as the likely final Hg complexing agent.

In another study employing AC for GEM measurements, Jeon et al. [6] deployed MerPAS® passive air samplers (PASs) containing sulfur impregnated AC along the northern Gulf of Mexico. The region has high Hg wet deposition rates and high levels of Hg in seafood compared to other coastlines in the USA. Unlike the Tekran speciation system discussed earlier, these PASs are relatively inexpensive and require no external power. Thus, they are typically deployed to increase area coverage and improve spatial resolution, albeit with poorer temporal resolution since they often need to be placed outdoors for several weeks to capture enough Hg for measurement. The PASs discriminate landscape and seasonal effects if given sufficient collection time, adequate analytical precision, and low blank levels [13]. Along the Gulf Coast, it was found that concentrations varied depending on proximity to sources and site characteristics, with a coastal marsh having the lowest GEM levels possibly due to uptake by vegetation. The PASs gave slightly lower concentrations compared to an active sampling system at Grand Bay, but generally showed similar seasonal patterns. Overall, the work demonstrates that PASs can provide insight into GEM levels and the factors affecting them along coastal regions.

As noted, three of the papers focused on Hg deposition. In Washington, USA, Beutel and colleagues [2] quantified both wet and dry Hg deposition at rural and suburban sites using a direct measurement approach with simple, low-cost equipment. The authors used an aerodynamic "wet sampler" for assessing dry deposition. GOM, PBM, and some GEM were collected in a thin layer of a recirculating acidic aqueous solution placed on a Teflon plate. The setup was operated for several days before the solution was collected, preserved, and analyzed for total-Hg. Wet deposition sampling was more standard. The rate was calculated as mass accumulation divided by the area of the collecting funnel and duration of the sampling event. The authors concluded that direct measurement approaches are useful in assessing temporal and spatial patterns of Hg deposition, and for comparing results to other approaches and estimates from numerical air quality models. They also show that agricultural burning in rural areas can lead to elevated levels of dry deposition and that short-term rain events account for ~20% of Hg deposition during the dry season.

Meanwhile, Tang et al. [4] examined Hg wet deposition on Chongming Island, China, from 2014–2018 to understand Hg wet deposition characteristics over multiple years. Notably, this is a period that saw the implementation of the Minamata Convention and anthropogenic Hg emission controls that allowed evaluation of the effectiveness of those controls. Volume-weighted mean Hg concentrations decreased during the study period, which was explained by decreasing atmospheric Hg levels and anthropogenic emission reductions. The authors also examined the impact of meteorological variations and showed that large-scale meteorological circulation events (e.g., monsoons) significantly impact Hg wet deposition. Thus, they recommend using long-term Hg wet deposition flux values in future Hg assessment programs to evaluate the impact of anthropogenic emissions reductions and inter-annual meteorological conditions.

Biomonitoring is yet another approach to measure Hg deposition, but one that requires no sampling equipment, other than that needed to retrieve the biota. Here, Stafilov et al. [3] used two species of moss blanketing North Macedonia to examine spatial and temporal patterns of Hg deposition within the eastern European region. The authors determined Hg in the samples collected as far back as 2002. Analysis of the median values showed an increase from 2002 to 2010 and a slight reduction from 2010 to 2015. Mercury distribution maps showed that sites with increased concentrations of Hg in moss were likely impacted by anthropogenic pollution; sites include thermoelectric power plants and a former chlor-alkali plant. It was concluded that Hg air pollution in the region is highest in industrialized areas. The authors rightfully note that such work is important for modeling Hg pollution and monitoring future trends in deposition, in order to assess and preserve ecosystem health. This is also a nice segue to the last two papers to be highlighted.

A good way to conclude this editorial are the contributions from Brown et al. [7] and Marshall et al. [8] on the environmental and health impacts of mining, bringing us back to the Minamata Convention on Mercury. It is undeniable that artisanal and small-scale gold mining (ASGM) employing Hg has major impacts on Hg pollution. Gold amalgamation with Hg^0, which incidentally is used in many of our instruments to concentrate and isolate Hg prior to spectroscopic analysis, is still commonly used in the developing world to separate gold from unwanted minerals, despite the means to do so without Hg [14]. I applaud the work of Brown and colleagues involving a training program to monitor elemental Hg emissions originating from gold shops in Guyana, South America. The effort included both locals and undergraduate student researchers from the USA, offering a meaningful educational experience for both. Several gold shops had measurements exceeding 100,000 ng/m^3, the guideline for occupational exposure limits in the USA. The authors note that while their work identified this significant source of Hg^0 emissions, it did not provide insight into the fate of this Hg. They also suggest that future mapping incorporating data from passive air samplers, like those discussed earlier, may add to our understanding of the fate of Hg^0 emitted from gold shops. Finally, they astutely point out that ASGM continues to be a global challenge faced by not only Guyana and other nations engaged in ASGM activities, but also all signatory nations of the Minamata Convention.

Last, but not least, Marshall et al. [8] reported on Hg challenges faced by Mexico involving regulatory, trade, and environmental impacts. While official Hg exports have declined, primary artisanal Hg mining in Mexico continues to proliferate, as does its associated problems. The authors include a brief history of Hg mining in Mexico, a description of the key stages of the mining process, and an analysis of Hg supply and trade in several countries of South America that have large ASGM sectors, leading to the examination of the regulatory control suggested by the Minamata Convention directives. They also provide atmospheric Hg concentrations measured at mine sites. The authors conclude that as the gold price remains high and artisanal gold mining using Hg amalgamation proliferates around the world, the demand for Hg from countries like Mexico and Indonesia will continue unabated, even with implementation of international agreements like the Minamata Convention. Finally, they discuss the economic alternatives that could be promoted in the region to substitute the destructive practices associated with primary Hg mining. I personally appreciate the final section on recommendations and future steps, which is a must read for stakeholders.

In summary, this group of articles provide a valuable update on atmospheric Hg research, showing not just how far we have come as a research community, but how far we must still go. The research stems from around the world, exemplifying that Hg is a global pollutant that affects us all. I thank the authors for their valuable contributions and hope this issue sparks some thought, collaboration, or simply serves as a resource to move us forward in a rapidly changing world and climate.

Funding: This research received no external funding.

Institutional Review Board Statement: Not applicable.

Informed Consent Statement: Not applicable.

Data Availability Statement: Not applicable.

Conflicts of Interest: The authors declare no conflict of interest.

References

1. Driscoll, C.T.; Mason, R.P.; Chan, H.M.; Jacob, D.J.; Pirrone, N. Mercury as a Global Pollutant: Sources, Pathways, and Effects. *Environ. Sci. Technol.* **2013**, *47*, 4967–4983. [CrossRef] [PubMed]
2. Beutel, M.W.; DeSilva, L.; Amegbletor, L. Direct Measurement of Mercury Deposition at Rural and Suburban Sites in Washington State, USA. *Atmosphere* **2021**, *12*, 35. [CrossRef]
3. Stafilov, T.; Barandovski, L.; Šajn, R.; Bačeva Andonovska, K. Atmospheric Mercury Deposition in Macedonia from 2002 to 2015 Determined Using the Moss Biomonitoring Technique. *Atmosphere* **2020**, *11*, 1379. [CrossRef]
4. Tang, Y.; Wu, Q.; Gao, W.; Wang, S.; Li, Z.; Liu, K.; Han, D. Impacts of Anthropogenic Emissions and Meteorological Variation on Hg Wet Deposition in Chongming, China. *Atmosphere* **2020**, *11*, 1301. [CrossRef]
5. Živković, I.; Berisha, S.; Kotnik, J.; Jagodic, M.; Horvat, M. Traceable Determination of Atmospheric Mercury Using Iodinated Activated Carbon Traps. *Atmosphere* **2020**, *11*, 780. [CrossRef]
6. Jeon, B.; Cizdziel, J.V.; Brewer, J.S.; Luke, W.T.; Cohen, M.D.; Ren, X.; Kelley, P. Gaseous Elemental Mercury Concentrations along the Northern Gulf of Mexico Using Passive Air Sampling, with a Comparison to Active Sampling. *Atmosphere* **2020**, *11*, 1034. [CrossRef]
7. Brown, S.T.; Bandoo, L.L.; Agard, S.S.; Thom, S.T.; Gilhuys, T.E.; Mudireddy, G.K.; Eechampati, A.V.; Hasan, K.M.; Loving, D.C.; Seney, C.S.; et al. A Collaborative Training Program to Assess Mercury Pollution from Gold Shops in Guyana's Artisanal and Small-Scale Gold Mining Sector. *Atmosphere* **2020**, *11*, 719. [CrossRef]
8. Marshall, B.G.; Camacho, A.A.; Jimenez, G.; Veiga, M.M. Mercury Challenges in Mexico: Regulatory, Trade and Environmental Impacts. *Atmosphere* **2020**, *12*, 57. [CrossRef]
9. Gustin, M.S.; Dunham-Cheatham, S.M.; Huang, J.; Lindberg, S.; Lyman, S.N. Development of an Understanding of Reactive Mercury in Ambient Air: A Review. *Atmosphere* **2021**, *12*, 73. [CrossRef]
10. Stratton, W.J.; Lindberg, S.E. Use of a refluxing mist chamber for measurement of gas-phase mercury(ii) species in the atmosphere. *Water Air Soil Pollut.* **1995**, *80*, 1269–1278. [CrossRef]
11. Ebinghaus, R.; Jennings, S.G.; Schroeder, W.H.; Berg, T.; Donaghy, T.; Guentzel, J.; Kenny, C.; Kock, H.H.; Kvietkus, K.; Landing, W.; et al. International field intercomparison measurements of atmospheric mercury species at Mace Head, Ireland. *Atmos. Environ.* **1999**, *33*, 3063–3073. [CrossRef]
12. Feng, X.B.; Sommar, J.; Gardfeldt, K.; Lindqvist, O. Improved determination of gaseous divalent mercury in ambient air using KCl coated denuders. *Fresenius J. Anal. Chem.* **2000**, *366*, 423–428. [CrossRef] [PubMed]
13. Jeon, B.; Cizdziel, J.V. Can the MerPAS Passive Air Sampler Discriminate Landscape, Seasonal, and Elevation Effects on Atmospheric Mercury? A Feasibility Study in Mississippi, USA. *Atmosphere* **2019**, *10*, 617. [CrossRef]
14. Drace, K.; Kiefer, A.M.; Veiga, M.M.; Williams, M.K.; Ascari, B.; Knapper, K.A.; Logan, K.M.; Breslin, V.M.; Skidmore, A.; Bolt, D.A.; et al. Mercury-free, small-scale artisanal gold mining in Mozambique: Utilization of magnets to isolate gold at clean tech mine. *Atmosphere* **2012**, *32*, 88–95. [CrossRef]

Review

Development of an Understanding of Reactive Mercury in Ambient Air: A Review

Mae Sexauer Gustin [1,*], Sarrah M. Dunham-Cheatham [1], Jiaoyan Huang [2], Steve Lindberg [3] and Seth N. Lyman [4]

1. Department of Natural Resources and Environmental Science, University of Nevada-Reno, Reno, NV 89557, USA; sdunhamcheatham@unr.edu
2. Sonoma Technology, Petaluma, CA 94952, USA; huangj1311@gmail.com
3. Emeritus Fellow Oak Ridge National Laboratories, Graeagle, CA 96103, USA; lindbergsteve6@gmail.com
4. Bringham Research Laboratory, Utah State University, Vernal, UT 84078, USA; seth.lyman@usu.edu
* Correspondence: mgustin@cabnr.unr.edu

Abstract: This review focuses on providing the history of measurement efforts to quantify and characterize the compounds of reactive mercury (RM), and the current status of measurement methods and knowledge. RM collectively represents gaseous oxidized mercury (GOM) and that bound to particles. The presence of RM was first recognized through measurement of coal-fired power plant emissions. Once discovered, researchers focused on developing methods for measuring RM in ambient air. First, tubular KCl-coated denuders were used for stack gas measurements, followed by mist chambers and annular denuders for ambient air measurements. For ~15 years, thermal desorption of an annular KCl denuder in the Tekran® speciation system was thought to be the gold standard for ambient GOM measurements. Research over the past ~10 years has shown that the KCl denuder does not collect GOM compounds with equal efficiency, and there are interferences with collection. Using a membrane-based system and an automated system—the Detector for Oxidized mercury System (DOHGS)—concentrations measured with the KCl denuder in the Tekran speciation system underestimate GOM concentrations by 1.3 to 13 times. Using nylon membranes it has been demonstrated that GOM/RM chemistry varies across space and time, and that this depends on the oxidant chemistry of the air. Future work should focus on development of better surfaces for collecting GOM/RM compounds, analytical methods to characterize GOM/RM chemistry, and high-resolution, calibrated measurement systems.

Keywords: cation exchange membrane; denuder; dual channel system; mist chamber; nylon membrane

1. Introduction

1.1. Discovery of GOM

Mercury (Hg) exists in the atmosphere as three forms: gaseous elemental (GEM), gaseous oxidized (GOM), and particulate-bound (PBM). Often, GOM and PBM concentrations are combined and collectively described as reactive Hg (RM). In the beginning, the atmospheric Hg research community focused on development of methods for GEM and did not know GOM existed. Now, GOM is known to be emitted from anthropogenic point sources and formed by atmospheric oxidation reactions of GEM with ozone (O_3), hydroxyl radical ($OH^·$), nitrate (NO_3), hydrogen peroxide (H_2O_2), and/or halogen-containing compounds ($Cl^·$, $Br^·$, ClO, BrO, $ClBr$) [1,2]. A more recent paper by Saiz-Lopez et al. [3] provides an update on current thinking regarding our understanding with respect to reactions and points out, using a global model based on bromine-induced GEM oxidation that other oxidation mechanisms are needed in the troposphere to explain observations.

In 1979, Fogg and Fitzgerald [4] postulated that since GEM is not highly water soluble, concentrations measured in precipitation could not be explained by GEM alone. Kothny (1973) [5] suggested Hg adsorbed to aerosols was the Hg form present in precipitation. Brosset (1983) [6], based on equilibrium coefficients developed by Iverfeldt, noted that

HgCl$_2$ and CH$_3$HgCl could explain observed concentrations. A mechanism for oxidation was proposed by Iverfeldt and Lindqvist [7] that entailed oxidation of GEM in water by ozone.

At the time, atmospheric Hg measurements were made using gold traps with a glass wool filter upstream to capture the particulate component. However, there were inconsistent results with the particulate filter. Research then focused on collecting GEM using gold surfaces such as gold-coated denuders [8,9] and gold-coated sand traps [10,11] Currently, gold-coated sand traps are the standard method for measurement of GEM, there is still controversy as to whether this is a measurement of GEM or total gaseous Hg (TGM). The Global Mercury Observation System standard operating procedure states that a soda lime trap in front of the Tekran 2537 removes GOM, though this has not been adequately tested.

In 1996, in a critical review paper on Hg speciation in flue gases associated with coal combustion, Galbreath and Zygarlicke [12] pointed out that a variety of RM compounds should exist, including Cl-, O-, and S-based compounds. They also reported Hg(II) (oxidized) forms did exist in the flue gas, based on measurements using USA Environmental Protection Agency (EPA) Method 29, EPA Method 101A, and the modified Method 101A and laboratory tests. Lindberg et al. [13,14] suggested that if such oxidized forms of gaseous Hg persisted in ambient air, they had the potential to be significant contributors to Hg deposition.

1.2. Early Development of Methods

In 1995, a landmark paper was published that described the use of a mist chamber method for measuring RM and provided the first measurements of RM in ambient air [15] A similar type of method had been attempted earlier by Brosset and Lord [16] using bubblers and long sampling times. Brosset and Lord [16] concluded that measured GOM was an artifact and better approaches were needed. The mist chamber used a single nebulizer nozzle, operated at a flow rate of 15 to 20 L min^{-1}, and collected samples in 20 mL of solution [15]. Stratton and Lindberg [15] reported that one-hour samples contained 50 to 200 pg RM. The mist chamber was deployed at two locations, Tennessee and Indiana, and concentrations of 50 to 150 pg m^{-3} were reported; similar trends were observed under field conditions at the two sites, leading to the conclusion that the method provided reasonable results [15]. The main concerns with this method were artifacts associated with O$_3$ and the presence of aerosols, which were extensively tested [15,17] Artifact formation was considered sufficiently slow relative to sampling times. Data collected using the mist chamber method was significantly correlated with temperature, solar radiation, O$_3$, SO$_2$, and total gaseous Hg [18]. Additional work using the sampling system further demonstrated the utility of the method and the limited effect of artifacts on the measurements [18]. Two known drawbacks of the system were that it was not calibrated, and potential for artifacts could vary by sampling location.

At this same time, researchers were also testing the use of membranes for both PBM and GOM capture. Ebinghaus et al. [19] applied Teflon disc filters, Whatman quartz filters, and quartz wool plugs, or Au traps preceded by Au denuders for PBM, and ion exchange membranes for GOM measurements. PBM measured by the different methods ranged from 5 to 100 pg m^{-3} with the highest concentrations observed on gold traps after a denuder. Ion exchange membranes measured concentrations of 40 to 95 pg m^{-3}, higher than denuder methods by 10 to 20 pg m^{-3} that were determined after liquid extraction. Munthe et al. [20] explored the use of microquartz fiber filters, cellular acetate, glass fiber, and Teflon filters for measurement of PBM; these results were quite variable.

Denuder methods for measuring GOM in ambient air were first pioneered by Oliver Lindqvist and his collaborators (e.g., Xiao et al. [8]; Feng et al. [21]). Their method utilized a KCl-coated tubular denuder, with GOM quantified using a liquid extraction. Comparison of the tubular and annular denuders showed similar recoveries in two studies in which the tubular denuder was liquid extracted and the annular denuder desorbed (Munthe et al. [20]

Nacht et al. [22]); Sommar et al. [23] reported lower GOM concentrations for annular denuders. In the Munthe et al. [20] intercomparison, mist chamber measurements were made, and concentrations agreed with those measured by the denuders. Nacht et al. [22] worked in a highly Hg contaminated location, reported RM concentrations of up to 75,000 pg m^{-3} with the highest values being above mine tailings.

In 2000, Steffen et al. [24] reported on the use of a cold regions pyrolysis unit manufactured by Tekran to allow for measurement of total gaseous mercury, while simultaneously measuring GEM. Their measurements were conducted during a Hg depletion event in the Arctic at Alert, Nunaurt, Canada. They observed that 48% of the converted GEM was measured as RM with the pyrolyzer unit and the rest deposited to snow.

Landis et al. [25] was the first to report on the use of an annular denuder in an automated system from which GOM could be thermally desorbed repeatedly to improve temporal resolution. During the period of denuder development, Landis et al. [25] and Xiao et al. [26] tested the efficiency of KCl denuders to collect permeated $HgCl_2$, with the latter testing CH_3HgCl as well. Neither study was conclusive; for example, Xiao et al. [26] utilized clean air, and the spiked GOM concentrations are not reported; while Landis et al. [25] data were limited (n = 2) and spike concentrations were one-to-two orders of magnitude higher than reported ambient concentrations (c.f. Valente et al. [27]). Feng et al. [21] reported limited laboratory tests of a tubular denuder loaded with hundreds to 1200 pg in three tests to determine breakthrough; however, the air used for the tests was not made clear. They used thermal desorption of the KCl denuders instead of liquid extraction The authors suggested that if a pyrolyzer was not used after desorption of the denuder that volatile or semi-volatile compounds trapped in the denuder would be released and deposit on the surface of the gold trap, risking passivation. Feng et al. [21] also recommended a denuder desorption temperature of 900 °C, due to the presence of a dual peak that they suggested was not a Hg compound, but volatile organic compounds that interfered with the analysis of Hg. No interference testing was reported in these studies. Feng et al. [21] commented on the fact that if water vapor condensed on the denuder, the sampling efficiency would decrease. Landis et al. [25] suggested that the temperature of the denuder be maintained at 50 °C to prevent hydrolysis of the KCl coating.

The Tekran® 2537/1130/1135 speciation system (Tekran system manufactured by Tekran, Toronto, Canada) was first introduced in 2002, and collects GEM, GOM, and PBM, respectively [25]. Ambient air entering the Tekran system first passes through an elutriator used to prevent coarse particles (>2.5 µm) from moving into the system; the flow rate of the system determines the particle cut size and must be routinely monitored and adjusted. Air then passes through the KCl denuder (1130 module, GOM capture) and subsequently through a quartz fiber filter (1135 module, PBM capture). Downstream of these modules is a pyrolyzer, packed with quartz chips, used to reduce GOM and PBM to GEM at predetermined intervals. Lastly, the air enters the 2537 module, which collects GEM by way of amalgamation on one of two gold-coated sand traps; the two traps are used to alternately collect and desorb Hg, allowing for continuous collection at 2.5 + min resolution (commonly 5 min). GEM is desorbed from the cartridges at 325 to 370 °C, then carried by argon to a quartz cell where Hg is quantified using cold vapor atomic fluorescence spectroscopy (CVAFS). The method detection limit for GEM is 0.1 ng m^{-3}. While GEM is being measured, GOM and PBM are collected over 1 to 2 h. These operationally defined fractions are then sequentially thermally desorbed at 550 and 700 °C for GOM and PBM, respectively. GOM and PBM concentrations are quantified in Hg-free air after three flushing cycles without heating (system blank check), then one cycle of pyrolyzer heating, three cycles for desorbing the particulate filter, three cycles for desorbing the denuder, and two flushing cycles without heating to allow the system to cool. Desorbed GOM and PBM compounds pass through the pyrolyzer and are measured as GEM by the 2537. A soda lime trap is typically installed inline directly upstream of the 2537 inlet to prolong the life of the gold traps and is changed monthly. Typically, the 2537 module is calibrated every 24 h using an internal GEM permeation source, and less regularly using manual injections from an external GEM permeation source. It is noteworthy that calibrated 2537 units sampling the same air can generate concentrations that are up to 28% different (c.f., Gustin et al. [28]).

2. Early Method Intercomparisons

An early method comparison at Mace Head, Ireland, compared GOM measurements collected with the tubular denuder, and analyzed by liquid extraction, with those collected using ion exchange membranes with a quartz fiber filter upstream. These results showed that the denuder collected more GOM [19]; however, the quartz fiber filter could have influenced the amount of GOM collected on the ion exchange membrane [8,29,30]. Sheu and Mason [31] compared GOM concentrations measured using KCl annular denuders and ion exchange membranes, and also found higher GOM concentrations measured by the denuders; however, once again a quartz fiber filter preceded the ion exchange membrane GOM can be reduced on a quartz fiber filter, especially in the presence of relative humidity (see discussion below). Additionally, comparison of GOM collected using KCl-coated quartz fiber filters with that collected by cation exchange membranes (CEM) showed less GOM collected by the quartz fiber filters [32].

Sheu and Mason [31] at the Chesapeake Biological Laboratory, located 80 km SSE of Washington, D.C., compared RM measurements from ambient air using KCl denuders the mist chamber, and membranes. Membranes consisted of two polytetrafluoroethylene (PTFE) filters (47 mm diameter, 0.45 µm pore) in front of three cation exchange membranes (CEM) (47 mm diameter) housed in a five-stage filter holder. Unfortunately, the membranes were preceded by a long sampling inlet that we now know allows for deposition of RM and reduction to GEM (c.f. [28]). RM concentrations were measured every 2 h for the mist chamber, and 6 to 24 h for the membranes. Membrane and mist chamber concentrations were similar, but were lower than the denuder that sampled for 24 h. Comparing all methods over 3 days showed membrane and denuder RM concentrations to be higher than for the mist chamber. Concentrations measured at this location were as high as 550 pg m^{-3} but typically were 20 to 100 pg Hg m^{-3}.

In a comparison of the mist chamber and annular KCl-coated denuder in Florida, the mist chamber reported 6.5 times higher RM concentrations relative to the denuder [25] The difference was assumed to be an artifact due to PBM or reactions of GEM with acid used in the chamber. This conclusion, based on the works of Lindberg and Stratton [17] and Stratton and Lindberg [18], was not true, and the mist chamber measurement was likely more reliable than originally thought given what we now know about the KC denuder collection efficiency (see below). Landis et al. [25] recommended that denuders should not collect GOM for more than 12 h without being purged to avoid decreases in sampling efficiency. It is noteworthy as part of National Atmospheric Deposition Program Atmospheric Mercury Network (NADP AMNet) protocol denuders are changed out every two weeks. During this study, denuder measurements were systematically sawtoothing, and it was not clear why, since denuders were changed every 2 to 4 h. Denuder measurements in this study showed GOM concentrations of 0 to 200 pg Hg m^{-3}. A major limitation of the mist chamber was that it required constant attention and significant care to avoid contamination [18]. Thus, the Tekran system that involved less maintenance and oversight, became the method of choice for many researchers and management agencies.

3. Work Pointing to Issues with the Tekran Speciation System

The Tekran system, like the mist chamber, had no field calibration for GOM or PBM and the behavior of denuders in ambient air was not fully explored prior to large-scale deployment in monitoring networks. Stratton and Lindberg [16] stated that denuders were under-sampling GOM. Weiss Penzias et al. [33] pointed out the data being collected by the instrument could not be fully explained. Others were concerned that a mass balance for air Hg concentrations could not be closed. For example, GOM concentrations increased when GEM concentrations decreased; however, GOM concentrations were not sufficient to account for GEM lost, and based on dry deposition rates, GOM concentrations should have increased. Choi et al. [34] pointed out that GOM measured by the KCl denuder could not fully explain GEM loss due to oxidizing processes. These claims are supported by ongoing work.

Lyman et al. [35] investigated the potential for an O_3 interference associated with the Tekran system denuder and found the collection efficiency of permeated $HgBr_2$ decreased by 12 to 30% at O_3 concentrations of 6 to 100 ppb. The authors suggested reduction on the denuder wall by way of the following reaction: $HgCl_2 + 2O_3 \rightarrow Hg^0 + 2O_2 + 2ClO$ ($\Delta Gr = -85$ kj mole^{-1}, reported in the Open Discussion of this paper). Their results also implied that longer O_3 exposure led to less GOM recovery (10 to 26%, and 29 to 55% reduction in recovery for 2.5 and 30 min exposure to 30 ppb O_3, respectively). Earlier work by Lynam and Keeler [30] noted that that the KCl denuder removes O_3 and was highly efficient at low concentrations (95% removal at 28 ppb), but decreased as O_3 concentrations increased (6% removal at 120 ppb).

At the same time, Swartzendruber et al. [36] reported TGM concentration data collected by a Tekran system with an upstream pyrolyzer sampling air from the marine boundary layer. Another Tekran system simultaneously measured Hg in ambient air that passed through KCl denuders, and it was assumed GOM was scrubbed by the denuder. These data were collected during five flights over the Pacific Northwest, USA. GOM concentrations measured by the Tekran system denuder were always lower than those calculated as the difference between the Tekran TGM and GEM measurements. The authors attributed this to a lack of recovery of GOM by the denuder.

4. Realization RM Was Not Being Accurately Measured

4.1. Surrogate Surface Data

In 2007 and 2009, Lyman et al. [37,38] presented work focused on development of a surrogate surface for measurement of dry deposition of GOM. The collection surface utilized was a cation exchange membrane (CEM), specifically supported ICE 450 membrane (Pall Corporation, P/N ICE45S3R), an acidic, negatively charged polysulfone CEM that selectively sorbed RM. The surrogate surface was deployed in an Aerohead dry deposition sampler, an aerodynamic polyoxymethylene disk (104 cm^2 surface area). The Aerohead is deployed downward-facing to minimizing collection of PBM and has a drip shield along the rim that prevents rain from impacting the membrane surface except during windy rain events or heavy downpours.

The Aerohead dry deposition method has been applied by others [39–41], is available commercially, and continues to be used by the United States Environmental Protection Agency (EPA) [42]. The samplers were deployed as part of an EPA initiative to develop a total maximum daily load for Hg in Florida. Peterson et al. [43] demonstrated that dry deposition estimates using a bi-directional atmospheric resistance model and Tekran GOM concentrations were lower than surrogate surface measurements of dry deposition at a site near Fort Lauderdale and Tampa but were more similar at Outlying Landing Field near Pensacola (Figure 1). Spatial trends observed in passive GOM concentrations, and Aerohead dry deposition measurements were different from the Tekran system data, leading to the conclusions that there were 1—atmospheric Hg forms not being measured by the Tekran system, and 2—different Hg compounds with different dry deposition velocities.

It is thought that given the design of the Aerohead, only GOM is collected. Thus, it is a measurement of dry deposition. However, the surface itself does not reflect natural systems, and measured deposition may be higher than is actually occurring. The CEM surrogate surface can be used to understand deposition to ecosystems with low canopy resistance, e.g., water, and it can be used to calibrate natural GOM dry deposition to natural systems.

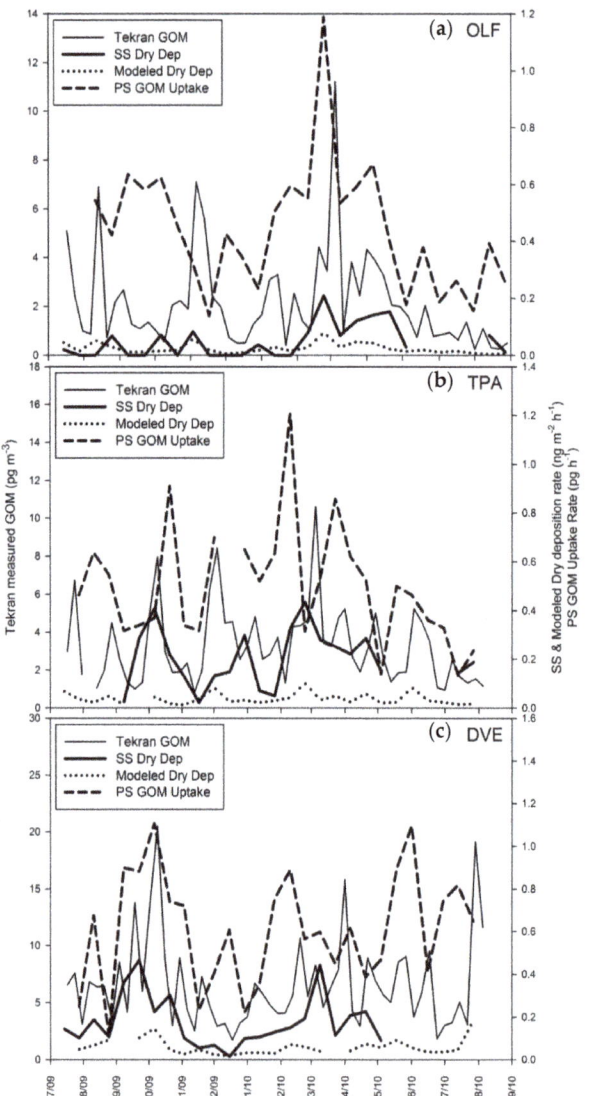

Figure 1. This figure shows data collected at three locations in Florida, USA: (**a**) Outlying Landing Field (OLF) near Pensacola, (**b**) Tampa (TPA), and (**c**) Davie (DVE) near Fort Lauderdale. Tekran system gaseous oxidized mercury (GOM) concentrations are presented, as are surrogate surface (SS) dry deposition measurements, modeled dry deposition using the Tekran system GOM data, and passive box samplers (PS) GOM uptake. From *Atmospheric Chemistry and Physics*, Peterson et al. [43] https://www.atmospheric-chemistry-and-physics.net/policies/licence_and_copyright.html.

4.2. RAMIX

The Reno Atmospheric Mercury Intercomparison eXperiment (RAMIX) took place from 22 August to 16 September 2012 [28]. The experiment focused on comparing Tekran system measurements with alternate methods for measurement of atmospheric Hg. A manifold was developed [44] that allowed for injection of $HgBr_2$, GEM, O_3, and water vapor into the air being sampled by each unit to calibrate instruments and test for interferences

Novel technologies and alternate methods tested during the comparison were the: University of Washington Detector for Oxidized Hg Species (DOHGS; [45]); University of Houston Mercury instrument (UHMERC); University of Miami Laser Induced Fluorescence (LIF; [46]); cavity ring-down spectroscopy system (Desert Research Institute); and nylon membranes. The UHMERC system measured only GEM and their data is reported in Gustin et al. [28]. The Desert Research Institute instrument did not collect any usable data during this experiment.

As show in Figure 2, during week 3, both Tekran systems, designated as Spec 1 (first in line in the manifold) and Spec 2, were sampling from the manifold. During week 4, Spec 2 was sampling ambient air at the site. It should be noted that Spec 2 concentrations were adjusted by 28% due to a consistent bias between the two Tekran 2537 modules GEM measurements.

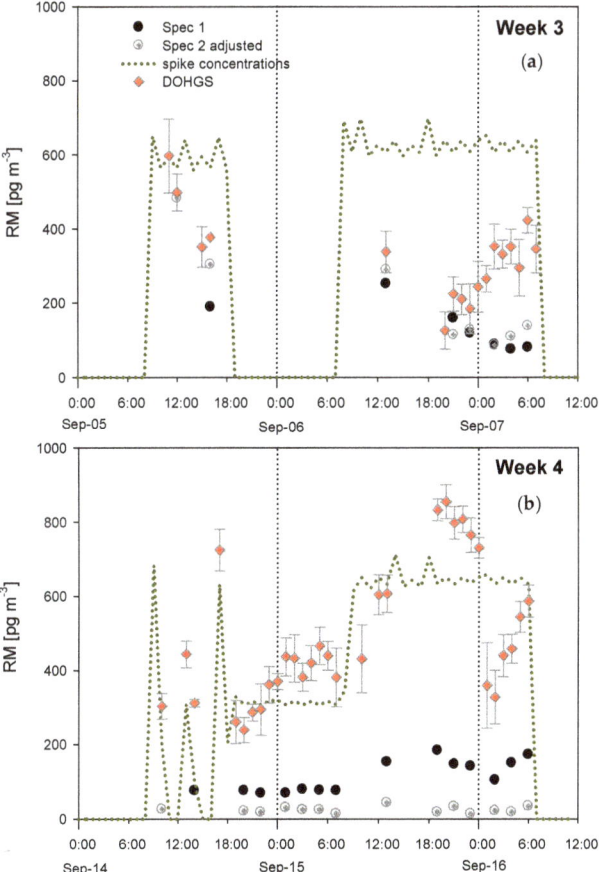

Figure 2. Hourly mean reactive mercury (RM) concentrations during Reno Atmospheric Mercury Intercomparison eXperiment (RAMIX) measured by two Tekran systems (Spec 1, Spec 2) and Detector for Oxidized mercury System (DOHGS) during $HgBr_2$ spikes. (**a**) Week 3; (**b**) Week 4. During week 3, both Tekran systems were sampling from the manifold, whereas in week 4, only Spec 1 was sampling from the manifold. Spec 1 and Spec 2 data represent a single hourly measurement, and the DOHGS data represent a 1-hour average of measurement made every 2.5 min. The error bars on DOHGS data represent 1σ. Reprinted (adapted) with permission from [28]. Copyright (2013) American Chemical Society.

DOHGS RM concentrations were higher than the Tekran RM measurements when sampling ambient air and during the spikes. At that time, the DOHGS used quartz wool to remove GOM from ambient air; during this experiment it was realized that when relative humidity increased the quartz wool lost GOM as GEM, and thus quartz wool has since been replaced by CEM for GOM collection (cf., 45). It is noteworthy that RM and $HgBr_2$ concentrations measured by the Tekran systems were typically higher for Spec 2 at night indicating that GOM was being generated in the manifold; this was hypothesized to be due to reactions with nitrogen compounds.

Several major conclusions resulted from this complicated method intercomparison. First, the Tekran system RM measurements were up to 13 times lower than those measured using the DOHGS. Second, the DOHGS was measuring a RM compound not being measured by the Tekran systems. Third, the DOHGS recovered 80% of the permeated $HgBr_2$; the lack of complete recovery could be explained by the loss of GOM from the quartz wool due to the presence of relative humidity [47]. Thus, quartz wool is not a good collection surface for GOM. Lastly, nylon membranes deployed in ambient air outside the manifold collected 30 to 50% more RM than the first Tekran system (Spec 1). The nylon measured concentrations during this experiment were an underestimate, given that nylon membranes consistently collect less RM than CEM that are currently thought to provide the more accurate RM measurements [48,49].

Concerns had previously been raised regarding potential artifacts associated with the Tekran system PBM measurement due to environmental temperatures and particle chemistry [50–52]. Data developed during RAMIX demonstrated that GOM was being collected on the particulate trap in the Tekran 1135 unit [28], supporting earlier suggestions of this possibility by Lynam et al. [30].

4.3. Additional Tests Following or Associated with RAMIX

Huang et al. [49] reported on a series of systematic laboratory tests that compared GOM uptake by KCl denuders with CEM and nylon membranes. Solid GOM compounds including $HgBr_2$, $HgCl_2$, and HgO, were used to permeate GOM into a laboratory manifold; additional compounds, $Hg(NO_3)_2$ and $HgSO_4$, were tested in Gustin et al [53]. Data collected using these three methods (Tekran system, nylon membrane, and CEM) were also collected in the field. A major finding of this work was that the polarizability of the compound influenced the ability of the denuder to collect the GOM compounds, with collection efficiency decreasing in the order: $HgBr_2 > HgCl_2 > HgO$. Moreover, in charcoal scrubbed air, GOM concentrations decreased in the order: CEM > nylon > KCl denuder. In tests comparing CEM versus nylon membrane GOM collection, the collection ratio of CEM:nylon was 1.5, 0.95, and 2.06 for $HgCl_2$, $HgBr_2$, and HgO, respectively. Similarly, the collection ratio of CEM:KCl denuder for the same compounds was 2.4, 1.5, and 3.7 respectively. Subsequent studies have demonstrated that, when comparing CEM data with a calibrated dual channel system, CEM are efficient at collecting GOM and RM compounds and the CEM is a good method for measurement of total GOM and RM compounds [54]. Nylon membranes do not collect GOM and RM compounds as well as the CEM [48,55].

The Huang et al. [49] paper also described the development of a thermal desorption system for determining GOM chemistry. The thermal desorption system was found to allow for the potential determination of chemistry of the RM compounds in ambient air using standard curves derived from the permeation of commercially available GOM compounds. Field data collected at three locations demonstrated that CEM > nylon > Tekran system GOM concentrations and the chemistry at each location varied, with -N and -S compounds collected at a location adjacent to a highway, halogenated compounds from the free troposphere collected at an agriculture-impacted location, and -N and -S compounds collected from the marine boundary layer.

Furthermore, building off the RAMIX project, Huang and Gustin [56] reported on a series of tests investigating the effect of relative humidity on KCl denuders, CEM, and nylon membranes. In these experiments, $HgBr_2$ was permeated into a manifold that had ports for membranes and denuders, along with relative humidity that was regulated between 25 to 75%. For the denuder, RM collection efficiency decreased to 60% when exposed to increasing levels of relative humidity, and when the humidity was removed the recovery was 60% of what it should have been. Denuders were also passivated over the two-week sampling time recommended by the NADP-AMnet. This effect was also observed during RAMIX. For the nylon membranes, RM collection decreased with increasing humidity and O_3 concentrations; increasing humidity increased RM collection on CEM.

McClure et al. [57] reported on measurements made at the North Birmingham South Eastern Aerosol Research and Characterization (SEARCH) site in summer 2013. This project focused on testing the performance of KCl denuders by permeating $HgBr_2$ into Hg- and O_3-free (clean) air, and ambient air. KCl denuders had ~95% collection efficiency in clean air, but the efficiency dropped to 20–54% in ambient air. Absolute humidity and O_3 were negatively correlated with $HgBr_2$ recovery. Follow up tests in a laboratory setting showed that increasing absolute humidity and O_3 resulted in the release of GEM from the denuder due to transformation of GOM to GEM [57].

5. Development of New Methods

5.1. Reactive Mercury Active System (RMAS)

The University of Nevada, Reno-Reactive Mercury Active System (RMAS) uses CEM and nylon membranes to actively collect RM from ambient air. CEM are used to measure concentrations, while nylon membranes allow for characterization of the chemistry. This system has been used to measure RM concentrations at a number of locations, including Nevada, Florida, Utah, Maryland, Hawaii, Sydney, Australia, the Southern Ocean, and Ny-Ålesund, Norway [48,58,59]. CEM have been used for similar purposes in different housings [60,61].

The RMAS has evolved over time and the current version, RMAS 2.0, is described in Luippold et al. [62]. Briefly, the system consists of six sampling ports, sampled using two vacuum pumps, with triplicate ports for nylon membranes and CEM. The membrane types are alternated in the system so if one pump goes offline data are still being collected. A critical flow orifice is used to regulate the flow through each sampling port to 1 Lpm. CEM are digested and analyzed using cold vapor atomic fluorescence following EPA Method 1631, and nylon membranes are thermally desorbed. The nylon desorption profiles are compared to standard profiles and the curves deconvoluted to determine the relative percent of each RM compound present on the membrane. Luippold et al. [48] concluded that the RM compounds measured on the nylon membranes using this method were reasonable given the atmospheric chemistry coming into the corresponding sampling location. Comparison of RM chemistry with measurement of anions F^-, Cl^-, Br^-, SO_4^{2-}, NO_2^-, NO_3^-, and PO_4^{3-} using an ion chromatograph also showed good agreement between RM chemistry and anion chemistry [48]. This work further demonstrated that the Tekran system denuder performs best in dry air with halogenated RM compounds. The best comparison between the KCl denuder and CEM measurements occurred on Moana Loa, Hawaii, and poor recoveries and large disagreements between the measurements existed at locations in Nevada and Maryland (Figure 3). The observed discrepancies between the Tekran system and membrane RM measurements in this study can be explained by the fact that the Tekran system denuder recoveries are influenced by the chemistry of RM compounds, where halogenated forms are collected more efficiently by the denuder relative to others (Table 1; [53]).

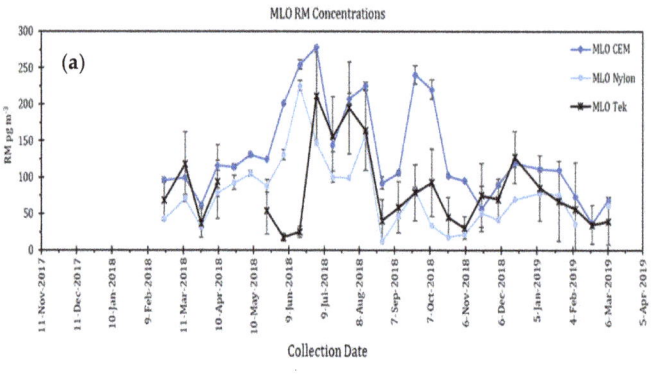
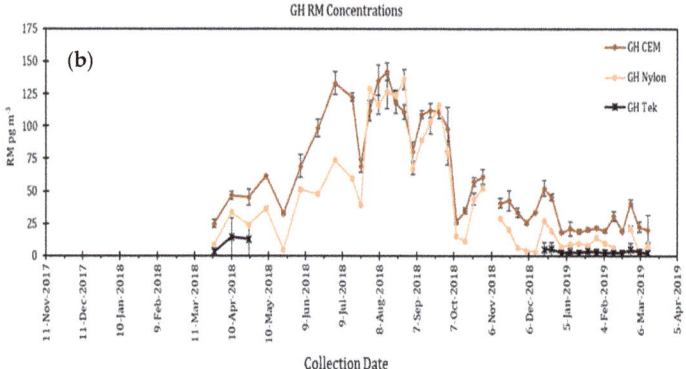
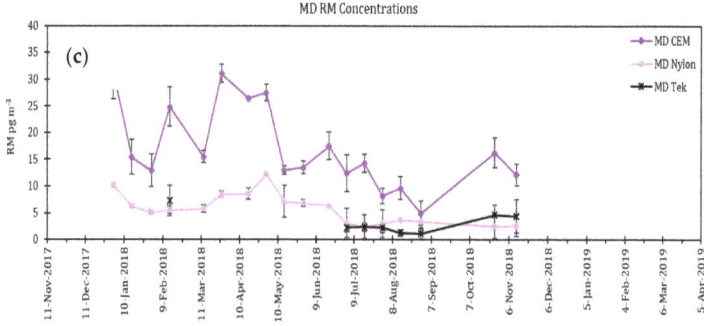

Figure 3. RM concentration data collected using Tekran system and RMAS 2.0 membranes (CEM and nylon) from November 2017 to March 2019 at 4 locations: Moana Loa, (MLO) Hawaii; Reno, NV (GH), and Piney Reservoir in Maryland (MD). (**a**) MLO; (**b**) GH (**c**) MD. Modified from Luippold et al. [48] Each panel shows CEM and nylon membrane RM concentrations as compared to Tekran RM data for MD and HI, and GOM data only for the Tekran deployed in NV.

Table 1. Regression equations comparing GOM concentrations measured by nylon membranes or cation exchange membranes (CEM) versus those measured by the Tekran system denuder. GOM permeations were performed using the UNR laboratory manifold system and charcoal-scrubbed air. From Gustin et al. [53]. https://www.atmospheric-chemistry-and-physics.net/policies/licence_and_copyright.html.

Comparison	$HgCl_2$	$HgBr_2$	HgO	$Hg(NO_3)_2$	$HgSO_4$
KCl denuder (x) vs. nylon membrane (y)	y = 1.6x + 0.002 r^2 = 0.97, n = 12	y = 1.7x + 0.01 r^2 = 0.99, n = 10	y = 1.8x + 0.02 r^2 = 0.99, n = 8	y = 1.4x + 0.04 r^2 = 0.90, n = 12	y = 1.9x − 0.1 r^2 = 0.6, n = 12
KCl denuder vs. CEM (y)	y = 2.4x + 0.1 r^2 = 0.58, n = 9	y = 1.6x + 0.2 r^2 = 0.86, n = 5	y = 3.7x + 0.1 r^2 = 0.99, n = 6	y = 12.6x − 0.02 r^2 = 0.50, n = 6	y = 2.3x + 0.01 r^2 = 0.95, n = 18

The RMAS has been further upgraded to include PTFE membranes upstream of two-in-line CEM and nylon membranes in three-stage filter packs. The PTFE membrane was added to allow for differentiating between PBM and GOM [55]. Figure 4 shows one set of data collected at the Nevada location with the PTFE membranes in one RMAS system, and no PTFE in a second RMAS. Hg concentrations on the nylon membranes with the upstream PTFE membrane were lower than the concentrations on the nylon membranes without the PTFE membrane. A few interesting observations from this work include that oxide compounds are found on both membranes, suggesting this form exists as both particulate and gaseous Hg compounds. Additionally, nitrogen and sulfur-based compounds were more likely to be associated with the aerosol phase; however, in some cases they were found on both nylon membranes.

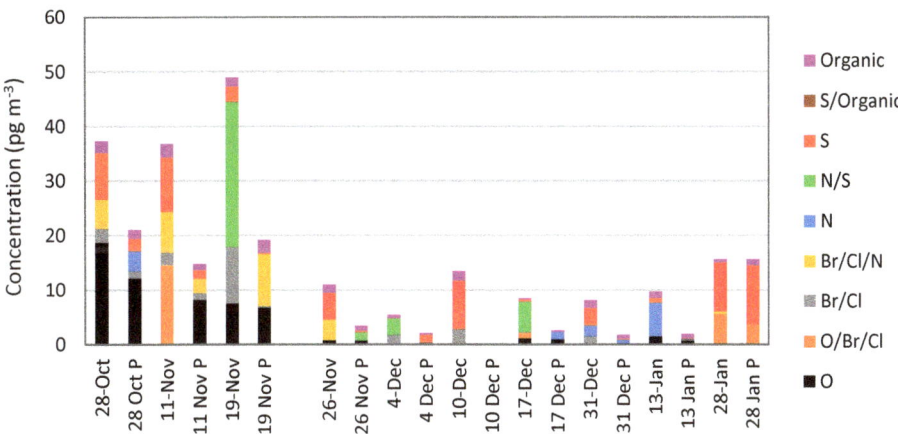

Figure 4. Breakdown of RM chemistries and concentrations measured on nylon membranes determined using thermal desorption and peak deconvolution. There are two sets of data for each sampling period; the date with the "P" is data derived from nylon membranes with an upstream polytetrafluoroethylene (PTFE) membrane. Compounds are designated as the major elements to which Hg(II) is bound based on deconvolution of thermal desorption profiles. Modified from Gustin et al. [53].

Data collected using the RMAS has demonstrated that GOM concentrations are much higher than previously thought, with concentrations up to 13 times higher than measured by the Tekran system, and that the chemical compounds of RM/GOM/PBM vary across space and time and are in-line with ancillary observations. This work reflects a step forward toward measurement of GOM and PBM concentrations and chemistry. Gustin et al. [55] further demonstrated the utility of the PTFE membrane as a means of discerning between GOM and PBM concentrations and chemistry. Concentrations measured using this system can be used to estimate dry deposition (Osterwalder et al. in progress). The mercury research community is actively seeking (via Mercury in the environment and links to deposition committee (MELD), a committee of the NADP) a measurement method or combination of methods to move forward our ability to measure Hg deposition as a community.

5.2. Dual Channel Systems

Dual channel systems provide an alternate automated method for measurement of RM. The first dual channel system was developed based on the operation of the University of Washington DOHGS (discussed above). The DOHGS system successfully measured elemental and oxidized Hg in several aircraft campaigns [47,63,64].

Its detection limit was ~100 pg Hg m^{-3}, limiting the utility at surface sites where RM concentrations tend to be lower. The DOHGS requires two Tekran 2537 modules, with perfectly matching precise calibrations, to collect data.

Gustin et al. [65] developed a dual channel system that requires only one Tekran 2537 module. This was an improvement over the need for two Tekran systems, for ease of use and cost effectiveness. This dual channel system consisted of a PTFE membrane at the inlet of the sampling system to remove particulates. Once air passed through this filter, the line was bifurcated into one line with a two-stage CEM to remove GOM, and second line with a pyrolyzer for TGM measurements. GOM was determined as the difference between the measurements between the two lines. Data were compared with a Tekran 2537/1130 system and two RMAS, one with CEM and nylon membranes, and the other with PTFE membranes upstream of the CEM and nylon membranes. The dual channel system and Tekran system underestimated GOM relative to the RMAS membrane measurements. For the dual channel system, the poor recovery was due to the long uncovered sampling line and relative humidity promoting reduction of GOM to GEM. The Tekran system exhibited low recovery due to the denuder.

A similar dual channel system was developed by Lyman et al. [66]. Their system differed from the Gustin et al. [65] dual channel system in a few key ways: the inlet line was covered (no light penetration), heated, and significantly longer; and the raw Tekran 2537 output was processed to calculate Hg concentrations based on peak height, and data were averaged to reduce the detection limit. This system resulted in a RM detection limit as low as ~15 pg Hg m^3 for a 1-h average; however, when operated at a different location the system detection limit was in the range of 40 pg Hg m^{-3}. Lyman et al. [66] showed that their dual channel system 1—could detect diurnal and other patterns in ambient GEM and GOM, 2—recovered 100% of HgBr$_2$ and HgCl$_2$ injected by an automated calibration system, 3—measured RM in the same range as the RMAS, and 4—accurately quantified RM concentrations at surface locations on an hourly timescale.

The collective work involving the RMAS and dual channel systems has provided a foundation for better understanding RM concentrations and chemistry, and information that modelers need to refine the biogeochemical Hg cycle.

5.3. Other Work Using CEM

Miller et al. [67] developed a method for measurement of GOM flux. The method utilized CEM to collect GOM from air interacting with mining based soils with a range of Hg contamination. They found that materials, such as those derived from mine tailing impoundments act as a direct emission source of RM compounds. This agrees with the work of Nacht et al. [22] who found high concentrations of RM over mine tailings in the field. The lower concentration mining-related substrates showed deposition of GOM with deposition velocities on the lower end of the range reported by Zhang et al. [68] in their review of Hg dry deposition.

Marusczak et al. [69] measured, using polyethersulfone membranes and a Tekran system, tropospheric RM at the Pic du Midi Observatory, France. They found the Tekran values to be systematically lower by a factor of 1.3 than the polyethersulfone membrane. They found a significant loss of RM (36%) from the denuder or inlet during flush blanks and if the flush blank was added to the Tekran RM measurement, the agreement with the

CEM was better (slope = 1.01, r^2 = 0.90), Concentrations measured in dry free tropospheric air were 198 ± 57 and 229 ± 58 pg m^{-3}), and agreed with in-flight observed RM, as well as model based estimates.

Field comparisons of the CEM and polyethersulfone membrane demonstrated that concentrations measured by the two are quite similar ([70]; unpublished data, Dr. Stefan Osterwalder, Université Grenoble Alpes). Preliminary results from a measurement campaign using the RMAS 2.0 and the Aerohead along with the Tekran system at Zeppelin Observatory on Svalbard, Norway, demonstrated that modeled deposition using Tekran data was lower than that determined using the Aerohead sampler and modeled values using RM concentrations measured by the RMAS system (Dr. Stefan Osterwalder, Université Grenoble Alpes; personal communication).

Size-resolved PBM measurements have been collected, usually using multi-stage impactors. This is reviewed and the biases discussed in Lyman et al. [54]. This review also discusses current understanding of reaction mechanisms.

5.4. Mass Spectrometric Methods

Several investigators have attempted to detect RM compounds using mass spectrometric techniques. Deeds et al. [71] pioneered this approach with preconcentration of Hg halide compounds on various surfaces, followed by thermal desorption into an atmospheric pressure chemical ionization mass spectrometer. With this method, the authors were unambiguously able to detect $HgBr_2$ and $HgCl_2$ in laboratory-generated samples. Measurements from samples collected from ambient air were less certain due to contaminants co-adsorbed during preconcentration.

Jones et al. [72] used cryogenic concentration, gas chromatography, and mass spectrometry to detect laboratory-generated and ambient GOM. Similar to Deeds et al. [71], they showed unequivocal detection of $HgBr_2$ and $HgCl_2$ in laboratory-generated samples, but they did not detect RM in ambient air samples. Gas chromatography resulted in the separation of RM compounds from each other and from contaminants, but non-halide Hg compounds failed to pass through the valves or the chromatography column, limiting the method applicability for some Hg compounds.

More recently, Khalizov et al. [73] used ion-drift chemical ionization mass spectrometry to detect $HgCl_2$. The authors did not analyze ambient samples, but they speculate that direct detection of RM compounds in the ambient atmosphere could be possible for future iterations of this method without prior preconcentration.

5.5. Oxidized Mercury Calibration Systems

Calibration of RM measurement systems with RM compounds has only occurred sporadically, though it has been repeatedly called for [54,74,75]. If a method for routine field calibration existed when the KCl denuder method first came into use, the low bias would likely have been quickly discovered, spurring the development of alternative methods at least a decade sooner.

Landis et al. [25] and Feng et al. [40] used $HgCl_2$ permeation tubes to test RM collection by KCl denuders. Landis et al. [25] found quantitative uptake, but their tests were in air scrubbed of ambient Hg and oxidants. Lyman et al. [35,62], and Finley et al. [44] during the RAMIX experiment, and Huang et al. [48], McClure et al. [57] and Huang and Gustin [56] used permeation tubes filled with various Hg (II) compounds to challenge KCl denuders and other measurement methods in ambient air, and, as discussed above, found denuder concentrations to be biased low. These early permeation tube-based methods were manual, meaning that the user had to manually turn a valve or connect tubing to add GOM to the sample air stream.

Lyman et al. [76] developed the first automated RM calibration system for ambient air measurements. They deployed the system at field sites and demonstrated that it delivered stable concentrations of $HgCl_2$ and $HgBr_2$ to measurement systems. Lyman et al. [66] expanded on this system by gravimetrically verifying Hg permeation rates from each permeation tube, though more work is needed to improve the gravimetric techniques. Lyman et al. [66] also showed that the automated calibration system can be used to quantify bias in RM measurements.

Sari et al. [77] tested two calibration systems that added $HgCl_2$ dissolved in ultrapure water to sample air streams. These systems generated RM concentrations much higher than is typical in ambient air, but evaporation methods like these are widely used for flue gas applications (e.g., Gonzalez et al. [78]), and could possibly be adapted for ambient measurements at lower concentrations.

While all of these discussed technologies will benefit from continued development and improvement, they show that routine, quantitative field calibration for oxidized Hg is necessary and possible. We advocate that all future measurements of oxidized Hg should include field calibration.

6. What We Have Learned

These new methods have repeatedly demonstrated that the KCl denuder is not adequate for measuring RM concentrations, and that new surfaces that can be used in an automated system are needed by the atmospheric Hg research community. We expect that KCl-denuder derived data is best for polar regions or high elevation locations in the free troposphere for these areas are dominated by halogenated compounds and dry air. The RMAS membrane-based system has been demonstrated to make accurate RM measurements when compared to a calibrated dual channel system [54]. Nylon membrane data have demonstrated that RM, GOM, and PBM chemistries vary across space and time, information critical for modelers trying to predict Hg deposition and the resulting impacts to ecosystems. However, the exact chemistry of RM compounds is not known, and this should be an emphasis of future work.

Although this work has been largely criticized due to the fact that it has shifted the paradigm away from the Tekran 1130 and 1135 units for being the industry standard for measurement of atmospheric RM, it has moved the Hg scientific community forward and led to better understanding of RM concentrations and chemistry.

7. Work Needed

The Hg research community is in need of a robust, high temporal resolution, calibrated method for measurement of GOM and PBM and/or RM concentrations and chemistry under all environmental conditions. In addition, methods developed should have clear quality control and quality assurance, there must be calibration standards, and tests will need to be done to achieve traceability to national standards that will need to be developed along with reference materials. [79] The RMAS has admitted limitations, including the long sampling duration required to collect sufficient Hg for analysis and the fact that the nylon membrane does not collect all forms of GOM/RM with equal efficiency. A new thermal desorption surface is needed, as the nylon membranes underestimate RM concentrations and have been demonstrated to collect less RM in the presence of water vapor and O_3. This surface will need be able to collect all compounds with equal efficiency and not have impacts due to air chemistry. It will be a challenge to find a surface that collects all compounds with equal efficiency and not have impacts due to air chemistry given the different forms have different water solubilities, particle affinities, and redox reactivity. Knowledge of the exact chemistry of RM compounds is still needed and has proven to be difficult to acquire using traditional analytical instruments due to the "stickiness" of RM compounds making it difficult to transport into a gas chromatography mass spectrometer [72]. More work is needed to develop a mass spectrometry method that will allow for identification of the chemistry. In the meantime, simply measuring GEM and TGM with a dual channel system using a pyrolyzer and CEM will allow for getting at GOM and a separate means will be needed for measurement of PBM. This method is a solution for those interested in deriving "real time" data, while the membrane system will be of use to those only interested in concentrations and obtaining an estimate of dry deposition. The latter would be useful in the face of the Minamata Convention. That said, the community has made significant progress over the past 30 years for measuring RM.

Author Contributions: Conceived the idea for the paper, M.S.G.; writing, M.S.G., S.M.D.-C., J.H., S.L. and S.N.L. All authors have read and agreed to the published version of the manuscript.

Funding: Major funding for the Gustin Laboratory's work on this topic came from the Electric Power Research Institute, US EPA, and the National Science Foundation.

Data Availability Statement: Data sharing not applicable.

Acknowledgments: The first author would like to thank those that blazed the trail for measuring GOM, as well as her many collaborators, students, and reviewers of her papers, for without the collective our progress would not have been possible. The authors thank the anonymous reviewers for their helpful comments.

Conflicts of Interest: The authors declare no conflict of interest.

References

1. Ariya, P.A.; Amyot, M.; Dastoor, A.; Deeds, D.; Feinberg, A.; Kos, G.; Poulain, A.; Ryjkov, A.; Semeniuk, K.; Subir, M.; et al. Mercury physicochemical and biogeochemical transformation in the atmosphere and at atmospheric interfaces: A review and future directions. *Chem. Rev.* **2015**, *115*, 3760–3802. [CrossRef] [PubMed]
2. Lin, C.-J.; Pehkonen, S.O. Aqueous phase reactions of mercury with free radicals and chlorine: Implications for atmospheric mercury chemistry. *Chemosphere* **1999**, *38*, 1253–1263. [CrossRef]
3. Saiz-Lopez, A.; Travnikov, O.; Sonke, J.E.; Thackray, C.P.; Jacob, D.J.; Carmona-García, J.; Francés-Monerris, A.; Roca-Sanjuán, R.; Acuña, A.U.; Dávalos, J.Z.; et al. Photochemistry of oxidized Hg(I) and Hg(II) species suggests missing mercury oxidation in the troposphere. *Proc. Natl. Acad. Sci. USA* **2020**, *117*, 30949–30956. [CrossRef]
4. Fogg, T.R.; Fitzgerald, W.F. Mercury in southern New-England coastal rains. *J. Geophys. Res. Oceans* **1979**, *84*, 6987–6989. [CrossRef]
5. Kothny, E.L. 3-phase equilibrium of mercury in nature. *Adv. Chem. Ser.* **1973**, *123*, 48–80.
6. Brosset, C. Transport of airborne mercury emitted by coal burning into aquatic systems. *Water Sci. Technol.* **1983**, *15*, 59–66. [CrossRef]
7. Iverfeldt, A.; Lindqvist, O. Atmospheric oxidation of elemental mercury by ozone in the aqueous phase. *Atmos. Environ.* **1986**, *20*, 1567–1573. [CrossRef]
8. Xiao, Z.F.; Munthe, J.; Lindqvist, O. Sampling and determination of gaseous and particulate mercury in the atmosphere using gold-coated denuders. *Water Air Soil Pollut.* **1991**, *56*, 141–151. [CrossRef]
9. Munthe, J.; Schroeder, W.H.; Xiao, Z.; Lindqvist, O. Removal of gaseous mercury from air using a gold coated denuder. *Atmos. Environ.* **1990**, *24*, 2271–2274.
10. Fitzgerald, W.F.; Gill, G.A. Sub-nanogram determination of mercury by 2-stage gold amalgamation and gas-phase detection applied to atmospheric analysis. *Anal. Chem.* **1979**, *51*, 1714–1720.
11. Bloom, N.; Fitzgerald, W.F. Determination of volatile mercury species at the picogram level by low-temperature gas-chromatography with cold-vapor atomic fluorescence detection. *Anal. Chim. Acta* **1988**, *208*, 151–161. [CrossRef]
12. Galbreath, K.C.; Zygarlicke, C.J. Mercury speciation in coal combustion and gasification flue gases. *Environ. Sci. Technol.* **1996**, *30*, 2421–2426. [CrossRef]
13. Lindberg, S.E.; Meyers, T.P.; Taylor, G.E.; Turner, R.R.; Schroeder, W.H. Atmosphere/surface exchange of mercury in a forest: Results of modeling and gradient approaches. *J. Geophys. Res.* **1992**, *97*, 2519–2528. [CrossRef]
14. Lindberg, S.E.; Turner, R.R.; Meyers, T.P.; Taylor, G.E.; Schroeder, W.H. Atmospheric concentrations and deposition of airborne mercury to Walker Branch Watershed. *Water Air Soil Pollut.* **1991**, *56*, 577–594. [CrossRef]
15. Stratton, W.J.; Lindberg, S.E. Use of a refluxing mist chamber for measurement of gas-phase mercury(ii) species in the atmosphere. *Water Air Soil Pollut.* **1995**, *80*, 1269–1278. [CrossRef]
16. Brosset, C.; Lord, E. Methylmercury in ambient air. Method of determination and some measurement results. *Water Air Soil Pollut.* **1995**, *82*, 739–750. [CrossRef]
17. Lindberg, S.E.; Stratton, W.J. Atmospheric mercury speciation: Concentrations and behavior of reactive gaseous mercury in ambient air. *Environ. Sci. Technol.* **1998**, *32*, 49–57. [CrossRef]
18. Stratton, W.J.; Lindberg, S.E.; Perry, C.J. Atmospheric mercury speciation: Laboratory and field evaluation of a mist chamber method for measuring reactive gaseous mercury. *Environ. Sci. Technol.* **2001**, *35*, 170–177. [CrossRef]
19. Ebinghaus, R.; Jennings, S.G.; Schroeder, W.H.; Berg, T.; Donaghy, T.; Guentzel, J.; Kenny, C.; Kock, H.H.; Kvietkus, K.; Landing, W.; et al. International field intercomparison measurements of atmospheric mercury species at Mace Head, Ireland. *Atmos. Environ.* **1999**, *33*, 3063–3073. [CrossRef]

20. Munthe, J.; Wängberg, I.; Pirrone, N.; Iverfeldt, Å.; Ferrara, R.; Ebinghaus, R.; Feng, X.; Gårdfeldt, K.; Keeler, G.; Lanzillotta, E. et al. Intercomparison of methods for sampling and analysis of atmospheric mercury species. *Atmos. Environ.* **2001**, *35*, 3007–3017. [CrossRef]
21. Feng, X.B.; Sommar, J.; Gardfeldt, K.; Lindqvist, O. Improved determination of gaseous divalent mercury in ambient air using KCl coated denuders. *Fresenius J. Anal. Chem.* **2000**, *366*, 423–428. [CrossRef] [PubMed]
22. Nacht, D.A.; Gustin, M.S.; Engle, M.A.; Zehner, R.Z.; Giglini, A.D. Quantifying total and reactive gaseous mercury at the Sulphur Banks Mercury Mine Superfund Site, Northern California. *Environ. Sci. Technol.* **2004**, *38*, 1977–1983. [CrossRef]
23. Steffen, A.; Schroeder, W.; Bottenheim, J.; Narayan, J.; Fuentes, J.D. Atmospheric mercury concentrations: Measurements and profiles near snow and ice surfaces in the Canadian Arctic during Alert 2000. *Atmos. Environ.* **2002**, *36*, 2653–2661. [CrossRef]
24. Sommar, J.; Feng, X.; Gardfeldt, K.; Lindqvist, O. Measurements of fractionated gaseous mercury concentrations over northwestern and central Europe, 1995–99. *J. Environ. Monit.* **1999**, *1*, 435–439. [CrossRef] [PubMed]
25. Landis, M.S.; Stevens, R.K.; Schaedlich, F.; Prestbo, E.M. Development and characterization of an annular denuder methodology for the measurement of divalent inorganic reactive gaseous mercury in ambient air. *Environ. Sci. Technol.* **2002**, *36*, 3000–3009. [CrossRef] [PubMed]
26. Xiao, Z.; Sommar, J.; Wei, S.; Lindqvist, O. Sampling and determination of gas phase divalent mercury in the air using a KCl coated denuder. *Fresenius J. Anal. Chem.* **1997**, *358*, 386–391. [CrossRef]
27. Valente, R.J.; Shea, C.; Humes, K.L.; Tanner, R.L. Atmospheric mercury in the Great Smoky Mountains compared to regional and global levels. *Atmos. Environ.* **2007**, *41*, 1861–1873. [CrossRef]
28. Gustin, M.S.; Huang, J.Y.; Miller, M.B.; Peterson, C.; Jaffe, D.A.; Ambrose, J.; Finley, B.D.; Lyman, S.N.; Call, K.; Talbot, R.; et al. Do we understand what the mercury speciation instruments are actually measuring? Results of RAMIX. *Environ. Sci. Technol.* **2013**, *47*, 7295–7306. [CrossRef]
29. Lu, J.Y.; Schroeder, W.H.; Berg, T.; Munthe, J.; Schneeberger, D.; Schaedlich, F. A device for sampling and determination of total particulate mercury in ambient air. *Anal. Chem.* **1998**, *70*, 2403–2408. [CrossRef]
30. Lynam, M.M.; Keeler, G.J. Artifacts associated with the measurement of particulate mercury in an urban environment: The influence of elevated ozone concentrations. *Atmos. Environ.* **2005**, *39*, 3081–3088. [CrossRef]
31. Sheu, G.-R.; Mason, R.P. An examination of methods for the measurements of reactive gaseous mercury in the atmosphere. *Environ. Sci. Technol.* **2001**, *35*, 1209–1216. [CrossRef] [PubMed]
32. Lyman, S.N.; Gustin, M.S. Determinants of atmospheric mercury concentrations in Reno, Nevada, U.S.A. *Sci. Total Environ.* **2009**, *408*, 431–438. [CrossRef] [PubMed]
33. Weiss-Penzias, P.; Jaffe, D.A.; McClintick, A.; Prestbo, E.M.; Landis, M.S. Gaseous elemental mercury in the marine boundary layer: Evidence for rapid removal in anthropogenic pollution. *Environ. Sci. Technol.* **2003**, *37*, 3755–3763. [CrossRef] [PubMed]
34. Choi, H.-D.; Huang, J.; Mondal, S.; Holsen, T.M. Variation in concentrations of three mercury (Hg) forms at a rural and a suburban site in New York State. *Sci. Total Environ.* **2013**, *448*, 96–106. [CrossRef] [PubMed]
35. Lyman, S.N.; Jaffe, D.A. Formation and fate of oxidized mercury in the upper troposphere and lower stratosphere. *Nat. Geosci.* **2012**, *5*, 114–117. [CrossRef]
36. Swartzendruber, P.C.; Jaffe, D.A.; Finley, B. Development and first results of an aircraft-based, high time resolution technique for gaseous elemental and reactive (oxidized) gaseous mercury. *Environ. Sci. Technol.* **2009**, *43*, 7484–7489. [CrossRef]
37. Lyman, S.N.; Gustin, M.S.; Prestbo, E.M.; Marsik, F.J. Estimation of dry deposition of atmospheric mercury in Nevada by direct and indirect methods. *Environ. Sci. Technol.* **2007**, *41*, 1970–1976. [CrossRef]
38. Lyman, S.N.; Gustin, M.S.; Prestbo, E.M.; Kilner, P.I.; Edgerton, E.; Hartsell, B. Testing and application of surrogate surfaces for understanding potential gaseous oxidized mercury dry deposition. *Environ. Sci. Technol.* **2009**, *43*, 6235–6241. [CrossRef]
39. Castro, M.S.; Moore, C.; Sherwell, J.; Brooks, S.B. Dry deposition of gaseous oxidized mercury in western Maryland. *Sci. Total Environ.* **2012**, *417*, 232–240. [CrossRef]
40. Fang, G.C.; Lin, Y.H.; Chang, C.Y. Use of mercury dry deposition samplers to quantify dry deposition of particulate-bound mercury and reactive gaseous mercury at a traffic sampling site. *Environ. Forensics* **2013**, *14*, 182–186. [CrossRef]
41. Sather, M.E.; Mukerjee, S.; Allen, K.L.; Smith, L.; Mathew, J.; Jackson, C.; Callison, R.; Scrapper, L.; Hathcoat, A.; Adam, J.; et al. Gaseous oxidized mercury dry deposition measurements in the southwestern USA: A comparison between Texas, eastern Oklahoma, and the Four Corners area. *Sci. World J.* **2014**, *2014*, 580723. [CrossRef] [PubMed]
42. Sather, M.E.; Mukerjee, S.; Smith, L.; Mathew, J.; Jackson, C.; Callison, R.; Scrapper, L.; Hathcoat, A.; Adam, J.; Keese, D.; et al. Gaseous oxidized mercury dry deposition measurements in the Four Corners area and eastern Oklahoma, U.S.A. *Atmos. Pollut. Res.* **2013**, *4*, 168–180. [CrossRef]
43. Peterson, C.; Alishahi, M.; Gustin, M.S. Testing the use of passive sampling systems for understanding air mercury concentrations and dry deposition across Florida, USA. *Sci. Total Environ.* **2012**, *424*, 297–307. [CrossRef] [PubMed]
44. Finley, B.D.; Jaffe, D.A.; Call, K.; Lyman, S.; Gustin, M.S.; Peterson, C.; Miller, M.; Lyman, T. Development, testing, and deployment of an air sampling manifold for spiking elemental and oxidized mercury during the Reno Atmospheric Mercury Intercomparison Experiment (RAMIX). *Environ. Sci. Technol.* **2013**, *47*, 7277–7284. [CrossRef]
45. Ambrose, J.L.; Lyman, S.N.; Huang, J.Y.; Gustin, M.S.; Jaffe, D.A. Fast time resolution oxidized mercury measurements during the Reno Atmospheric Mercury Intercomparison Experiment (RAMIX). *Environ. Sci. Technol.* **2013**, *47*, 7285–7294. [CrossRef]

6. Hynes, A.J.; Everhart, S.; Bauer, D.; Remeika, J.; Ernest, C.T. In situ and denuder-based measurements of elemental and reactive gaseous mercury with analysis by laser-induced fluorescence results from the Reno Atmospheric Mercury Intercomparison Experiment. *Atmos. Chem. Phys.* **2017**, *17*, 465–483. [CrossRef]
7. Ambrose, J.L.; Gratz, L.E.; Jaffe, D.A.; Campos, T.; Flocke, F.M.; Knapp, D.J.; Stechman, D.M.; Stell, M.; Weinheimer, A.J.; Cantrell, C.A.; et al. Mercury emission ratios from coal-fired power plants in the southeastern United States during NOMADSS. *Environ. Sci. Technol.* **2015**, *49*, 10389–10397. [CrossRef]
8. Luippold, A.; Gustin, M.S.; Dunham-Cheatham, S.M.; Castro, M.; Luke, W.; Lyman, S.; Zhang, L. Use of multiple lines of evidence to understand reactive mercury concentrations and chemistry in Hawai'i, Nevada, Maryland, and Utah, USA. *Environ. Sci. Technol.* **2020**, *54*, 7922–7931. [CrossRef]
9. Huang, J.Y.; Miller, M.B.; Weiss-Penzias, P.; Gustin, M.S. Comparison of gaseous oxidized Hg measured by KCl-coated denuders, and nylon and cation exchange membranes. *Environ. Sci. Technol.* **2013**, *47*, 7307–7316. [CrossRef]
10. Rutter, A.P.; Schauer, J.J. The impact of aerosol composition on the particle to gas partitioning of reactive mercury. *Environ. Sci. Technol.* **2007**, *41*, 3934–3939. [CrossRef]
11. Rutter, A.P.; Schauer, J.J. The effect of temperature on the gas-particle partitioning of reactive mercury in atmospheric aerosols. *Atmos. Environ.* **2007**, *41*, 8647–8657. [CrossRef]
12. Talbot, R.; Mao, H.; Feddersen, D.; Smith, M.; Kim, S.Y.; Sive, B.; Haase, K.; Ambrose, J.; Zhou, Y.; Russo, R. Assessment of particulate mercury measured with the manual and automated methods. *Atmosphere* **2010**, *2*, 1–20. [CrossRef]
13. Gustin, M.S.; Amos, H.M.; Huang, J.; Miller, M.B.; Heidecorn, K. Measuring and modeling mercury in the atmosphere: A critical review. *Atmos. Chem. Phys.* **2015**, *15*, 5697–5713. [CrossRef]
14. Lyman, S.N.; Cheng, I.; Gratz, L.E.; Weiss-Penzias, P.; Zhang, L. An updated review of atmospheric mercury. *Sci. Total Environ.* **2020**, *707*, 135575. [CrossRef] [PubMed]
15. Gustin, M.S.; Dunham-Cheatham, S.M.; Zhang, L.; Lyman, S.; Choma, N.; Castro, M. Use of membranes and detailed HYSPLIT analyses to understand atmospheric particulate, gaseous oxidized, and reactive mercury chemistry. *Environ. Sci. Technol.* **2020**. [CrossRef]
16. Huang, J.Y.; Gustin, M.S. Uncertainties of gaseous oxidized mercury measurements using KCl-coated denuders, cation-exchange membranes, and nylon membranes: Humidity influences. *Environ. Sci. Technol.* **2015**, *49*, 6102–6108. [CrossRef] [PubMed]
17. McClure, C.D.; Jaffe, D.A.; Edgerton, E.S. Evaluation of the KCl denuder method for gaseous oxidized mercury using HgBr2 at an in-service AMNet site. *Environ. Sci. Technol.* **2014**, *48*, 11437–11444. [CrossRef]
18. Huang, J.Y.; Miller, M.B.; Edgerton, E.; Gustin, M.S. Deciphering potential chemical compounds of gaseous oxidized mercury in Florida, USA. *Atmos. Chem. Phys.* **2017**, *17*, 1689–1698. [CrossRef]
19. Gustin, M.S.; Pierce, A.M.; Huang, J.Y.; Miller, M.B.; Holmes, H.A.; Loria-Salazar, S.M. Evidence for different reactive Hg sources and chemical compounds at adjacent valley and high elevation locations. *Environ. Sci. Technol.* **2016**, *50*, 12225–12231. [CrossRef]
20. He, Y.; Mason, R.P. Comparison of reactive gaseous mercury measured by KCl-coated denuders and cation exchange membranes during the Pacific GEOTRACES GP15 expedition. *Atmos. Environ.* **2020**, *241*, 117973. [CrossRef]
21. Miller, M.B.; Howard, D.A.; Pierce, A.M.; Cook, K.; Keywood, M.; Powell, J.; Gustin, M.S.; Edwards, G.C. Atmospheric reactive mercury concentrations in coastal Australia and the Southern Ocean. *Sci. Total Environ.* **2021**, *751*, 141681. [CrossRef] [PubMed]
22. Luippold, A.; Gustin, M.S.; Dunham-Cheatham, S.M.; Zhang, L. Improvement of quantification and identification of atmospheric reactive mercury. *Atmos. Environ.* **2020**, *224*, 117307. [CrossRef]
23. Lyman, S.N.; Jaffe, D.A.; Gustin, M.S. Release of mercury halides from KCl denuders in the presence of ozone. *Atmos. Chem. Phys.* **2010**, *10*, 8197–8204. [CrossRef]
24. Gratz, L.E.; Ambrose, J.L.; Jaffe, D.A.; Shah, V.; Jaeglé, L.; Stutz, J.; Festa, J.; Spolaor, M.; Tsai, C.; Selin, N.E.; et al. Oxidation of mercury by bromine in the subtropical Pacific free troposphere. *Geophys. Res. Lett.* **2015**, *42*, 10–494. [CrossRef]
25. Gustin, M.S.; Dunham-Cheatham, S.M.; Zhang, L. Comparison of 4 methods for measurement of reactive, gaseous oxidized, and particulate bound mercury. *Environ. Sci. Technol.* **2019**, *53*, 14489–14495. [CrossRef]
26. Lyman, S.N.; Gratz, L.; Dunham-Cheatham, S.M.; Gustin, M.S.; Luippold, A. Improvements to the accuracy of atmospheric oxidized mercury measurements. *Environ. Sci. Technol.* **2020**, *54*, 13379–13388. [CrossRef]
27. Miller, M.B.; Gustin, M.S.; Edwards, G.C. Reactive mercury flux measurements using cation exchange membranes. *Atmos. Meas. Tech. Discuss.* **2018**. [CrossRef]
28. Zhang, L.; Wright, L.P.; Blanchard, P. A review of current knowledge concerning dry deposition of atmospheric mercury. *Atmos. Environ.* **2009**, *43*, 5853–5864. [CrossRef]
29. Marusczak, N.; Sonke, J.E.; Fu, X.W.; Jiskra, M. Tropospheric GOM at the Pic du Midi Observatory-Correcting bias in denuder based observations. *Environ. Sci. Technol.* **2017**, *51*, 863–869. [CrossRef]
30. Dunham-Cheatham, S.M.; Lyman, S.; Gustin, M.S. Evaluation of sorption surface materials for reactive mercury compounds. *Atmos. Environ.* **2020**, *242*, 117836. [CrossRef]
31. Deeds, D.A.; Ghoshdastidar, A.; Raofie, F.; Guérette, E.A.; Tessier, A.; Ariya, P.A. Development of a particle-trap preconcentration-soft ionization mass spectrometric technique for the quantification of mercury halides in air. *Anal. Chem.* **2015**, *87*, 5109–5116. [CrossRef] [PubMed]
32. Jones, C.P.; Lyman, S.N.; Jaffe, D.A.; Allen, T.; O'Neil, T.L. Detection and quantification of gas-phase oxidized mercury compounds by GC/MS. *Atmos. Meas. Tech.* **2016**, *9*, 2195–2205. [CrossRef]

73. Khalizov, A.F.; Guzman, F.J.; Cooper, M.; Mao, N.; Antley, J.; Bozzelli, J. Direct detection of gas-phase mercuric chloride by ion drift-chemical ionization mass spectrometry. *Atmos. Environ.* **2020**, *238*, 117687. [CrossRef]
74. Gustin, M.S.; Jaffe, D. Reducing the uncertainty in measurement and understanding of mercury in the atmosphere. *Environ. Sci. Technol.* **2010**, *44*, 2222–2227. [CrossRef]
75. Jaffe, D.A.; Lyman, S.; Amos, H.M.; Gustin, M.S.; Huang, J.; Selin, N.E.; Levin, L.; Ter Schure, A.; Mason, R.P.; Talbot, R.; et al. Progress on understanding atmospheric mercury hampered by uncertain measurements. *Environ. Sci. Technol.* **2014**, *48*, 7204–7206. [CrossRef]
76. Lyman, S.N.; Jones, C.; O'Neil, T.; Allen, T.; Miller, M.; Gustin, M.S.; Pierce, A.M.; Luke, W.; Ren, X.; Kelley, P. Automated calibration of atmospheric oxidized mercury measurements. *Environ. Sci. Technol.* **2016**, *50*, 12921–12927. [CrossRef]
77. Sari, S.; Timo, R.; Jussi, H.; Panu, H. Dynamic calibration method for reactive gases. *Meas. Sci. Technol.* **2019**, *31*, 034001. [CrossRef]
78. González, R.O.; Díaz-Somoano, M.; Antón, M.L.; Martínez-Tarazona, M.R. Effect of adding aluminum salts to wet FGD systems upon the stabilization of mercury. *Fuel* **2012**, *96*, 568–571. [CrossRef]
79. ISO/IEC Guide 98-3:2008. Available online: http://www.iso.org/sites/JCGM/JCGM-introduction.htm (accessed on 4 January 2021).

Article

Traceable Determination of Atmospheric Mercury Using Iodinated Activated Carbon Traps

Igor Živković [1], Sabina Berisha [2], Jože Kotnik [1,2], Marta Jagodic [1,2] and Milena Horvat [1,2,*]

[1] Department of Environmental Sciences, Jozef Stefan Institute, Jamova cesta 39, 1000 Ljubljana, Slovenia; igor.zivkovic@ijs.si (I.Ž.); joze.kotnik@ijs.si (J.K.); marta.jagodic@ijs.si (M.J.)
[2] Jozef Stefan International Postgraduate School, Jamova cesta 39, 1000 Ljubljana, Slovenia; sabina.berisha@ijs.si
* Correspondence: milena.horvat@ijs.si

Received: 4 June 2020; Accepted: 22 July 2020; Published: 24 July 2020

Abstract: Traceable determination of atmospheric mercury (Hg) represents a major analytical problem due to low environmental concentrations. Although Hg pre-concentration on activated carbon (AC) traps is a simple method for sample collection, Hg determination is difficult due to a complex matrix that cannot be easily digested using wet chemistry. Two approaches for Hg loading on iodinated AC, the purging of elemental mercury (Hg^0) and the spiking a solution of standard reference material (SRM), were used to test whether spiking SRM solution on AC can be used for the traceable determination of atmospheric mercury collected as Hg^0. Mercury on AC was determined using atomic absorption spectrometry after sample combustion. The detector's response for both loading methods was identical in a wide concentration range, indicating that the spiking of SRM on AC can, indeed, be used for the calibration of analytical systems used for the determination of atmospheric mercury. This was confirmed by the determination of Hg in a real atmospheric sample collected on an iodinated AC trap and using an SRM spiking calibration. Different ACs were compared regarding their ability to quantitatively capture Hg while having the lowest breakthrough. Use of a specific impregnating solution probably converted Hg on AC to Millon's iodide, as estimated from the fractionation thermogram.

Keywords: atmospheric mercury; traceable determination; iodinated activated carbon; thermal fractionation

1. Introduction

Toxic mercury (Hg) compounds can cause adverse effects on human health [1,2]. Recent modeling suggests that mercury in the atmosphere has increased three to six-fold compared to natural levels, mainly due to anthropogenic Hg emissions [3,4]. Hg emitted to the atmosphere has a long residence time (up to a year) and can travel long distances before being deposited to land or ocean surfaces [5]. One of topics in the scientific community, driven by the Minamata convention on mercury, is the quantification of the extent to which anthropogenic mercury emitted to the atmosphere is converted to inorganic mercury and subsequently methylated and incorporated in biota [6,7].

Total airborne mercury (TAM) consists of particulate bound mercury (PBM) and total gaseous mercury (TGM) [8]. TGM represents the sum of gaseous oxidized mercury (GOM, Hg(II)) and gaseous elemental mercury (GEM, Hg^0) [5]. Even though Hg concentrations in the atmosphere are elevated due to anthropogenic emissions, these values all are still exceptionally low, which represents a major analytical challenge. This is especially important when performing mercury speciation in the atmospheric samples because GOM and PBM are present at pg m^{-3}, while GEM is at ng m^{-3} levels [9,10]. Although there are reliable instruments capable of measuring GEM, Hg speciation still represents a major analytical challenge.

Pre-concentration of mercury is usually required for the determination of GEM with the exception of atomic absorbance spectrometers (AAS) with Zeeman background correction [11]. However, for some characteristic analysis, e.g., Hg speciation or the determination of mercury stable isotope ratios in atmospheric samples, mercury pre-concentration is an essential step [12]. The determination of mercury stable isotope ratios requires a considerable amount of mercury, usually more than 10 ng [12,13]. This requires a mercury pre-concentration system that can quantitatively collect all mercury present in the sample. Quantitative trapping is required to ensure that the fractionation of mercury stable isotopes does not occur during the pre-concentration step.

The two most commonly used methods for the pre-concentration of mercury are collection on gold-coated quartz/glass sand/beads or on impregnated activated carbon (AC) traps [12,14]. Although gold traps are commonly used for the determination of mercury in various instrumental setups, their application for prolonged collection time of mercury is questionable. Gold traps suffer from passivation of gold [15], especially under the influence of seawater aerosol due to the corrosive properties of chlorides. On the contrary, impregnated AC is rather durable and can also be used for the pre-concentration of mercury from ambient air or even from stack emissions and flue gases that are characterized by the presence of highly corrosive gases [16–19]. AC possesses a high adsorption capacity for mercury compounds and might, therefore, be used as an effective sorbent in analytical traps. Impregnated AC, particularly iodinated, brominated or chlorinated AC, has a particularly high affinity for mercury compounds [16,17]. Therefore, impregnated AC is used as an efficient adsorbent for the quantitative capturing of atmospheric mercury. The main drawback of AC is higher blanks compared to the gold traps, which requires the collection of considerable amounts of the sample to diminish the influence of these blanks on reliable Hg determination.

According to the US Environmental Protection Agency (EPA) Method 30B, AC traps are used for the collection of both GOM and GEM fractions [20]. This method is intended for use only under relatively low particulate conditions (e.g., sampling of Hg emissions from coal-fired combustion sources after all pollution control devices). Therefore, if appropriate filters are not used to remove particulate matter, PBM will also be collected on AC traps and the obtained results will represent TAM. As mercury in the atmosphere is mostly present as Hg^0, this is the fraction that is mostly collected on AC traps, alongside GOM and PBM.

It is of utter importance to perform a traceable determination of mercury using appropriate calibration. Appropriate calibration for GEM would be by Hg^0 so that the sample and the standard are present in the same oxidation state of mercury. Traceable calibration sources for GOM at low atmospheric concentrations are still under development [21]. The standard for Hg^0 is usually obtained from a bell–jar apparatus held at a certain temperature from which a known volume of headspace vapors is taken using a gas-tight syringe. The amount of mercury taken by syringe at a certain temperature is calculated using empirical Dumarey or Huber equations [22–24]. The utilization of a calibration system based on bell–jar apparatus might represent an error in the determination of mercury on an impregnated AC trap due to the strong temperature dependence of these equations, and consequently cause differences compared to other calibration techniques [25,26].

The second approach to the calibration of an analytical system is the use of Hg^0 produced by the reduction of standard reference material (SRM) NIST 3133 [27]. This approach is more appropriate because mercury determinations are traceable to this standard reference material. Nevertheless, this sort of calibration is somewhat tedious and time consuming as quantitative reduction of Hg(II) from NIST 3133 to elemental form (Hg^0) using tin(II) chloride requires a certain amount of time. In addition, impurities in reagents can cause blanks or contamination.

The alternative to this approach would be directly spiking the diluted SRM NIST 3133 onto a sample matrix [28]. However, difficulties can also arise consequently. Hg species are not the same in NIST 3133 standard solution (where Hg is present as Hg(II)) and in purged Hg^0 that is produced by the reduction of the former. The use of Hg(II) in solution for the calibration of system used for the determination of Hg^0 on AC is questionable due to the quite different chemical and physical properties of these two Hg species. The chemistry of their corresponding adsorption onto the impregnated AC is probably represented by different physisorption/chemisorption mechanisms. Calibration using standard reference material of soils and sediments was recently compared with the calibration using the spiking of NIST 3133 standard solution onto the sample matrix [28]. Although both calibration curves were linear, the detector's responses for the two calibration curves were not identical. This showed that the liberation of mercury from the matrix is strongly dependent on the mercury species and how well this species is bound to matrix [28]. Therefore, it is necessary to verify that this approach to calibration is, indeed, traceable to standard reference material before it can be used for the determination of mercury in atmospheric samples.

The objective of this work has been to test whether spiking of NIST 3133 standard reference material directly onto iodinated AC can be used for the calibration of system used for the traceable determination of mercury purged onto this trap as Hg^0. Two approaches for loading Hg onto AC were compared; the first approach was the purging of NIST 3133-traceable Hg^0 on iodinated AC traps, while the other one was the direct spiking of diluted NIST 3133 reference material onto the AC. The method was tested by the determination of a real atmospheric sample from a contaminated indoor site and calibrating using a spiking method. To estimate which AC has the best quantitative capturing of Hg^0 and the lowest breakthrough, the performance of several in-house prepared iodinated ACs were compared to that of a commercially available AC. In this work, a NH_4I-impregnated AC was selected because iodine forms much more stable Hg complexes compared to other halogens. In addition, ammonium ions convert Hg-iodide complexes to insoluble Millon's iodide. We wanted to test whether the polymeric structure of Millon's iodide is more stable than Hg-iodide complexes. Therefore, NH_4I-impregnated ACs were used instead of commonly used KI-impregnated or brominated/chlorinated ACs. Peak deconvolution of fractionation thermograms was performed and results were compared to literature to estimate the stability of mercury species present on AC traps.

2. Experiments

2.1. Preparation of In-House Impregnated Activated Carbon

In-house impregnated iodinated AC was prepared using a modified method following the study by Fu et al. [12]. Virgin AC (grain size 0.5–1.0 mm; Merck, Darmstadt, Germany) was cleaned by heating at 500 °C for five hours in the stream of nitrogen gas (0.2 L min^{-1}). After the virgin AC was cooled down to room temperature, 5.0 g was conditioned for 48 h in 0.5 L of impregnating solution containing known concentrations of impregnating salts. Used impregnating solutions were 0.10 mol L^{-1} NH_4I, 0.03 mol L^{-1} NH_4I, and 0.01 mol L^{-1} NH_4I (ACS grade, Sigma Aldrich, St. Louis, MO, USA).

Following equilibration in the impregnating solution, AC was rinsed three times with Type I purified water (electrical resistivity 18.2 MΩ cm; Milli-Q water, Merck, Darmstadt, Germany) and dried in rotavapor at 60 °C under reduced pressure (50 mbar) until dry. Impregnated iodinated AC was stored in an amber glass bottle prior to use. The general scheme for the preparation of the impregnated AC is presented in Scheme 1.

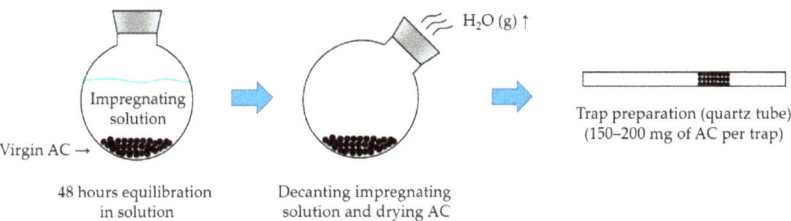

Scheme 1. General scheme for the preparation of iodinated activated carbon (AC) traps.

The amount of iodine bound on AC was estimated from the difference between the amount of iodine in the impregnating solution (0.01 mol L^{-1} NH$_4$I) prior to and after the equilibration of AC. The concentration of iodine in the impregnating solution was determined using ICP-MS after filtration through a 0.2-µm filter [29,30].

2.2. Preparation of Activated Carbon Traps

AC traps were prepared by filling a known amount (150–200 mg) of commercially available iodinated AC (AIC-500; APEX Instruments: Fuquay-Varina, NC, USA) or in-house prepared impregnated iodinated AC into a pre-cleaned quartz tube and fixed using a quartz wool. Quartz tube was pre-cleaned by washing with Milli-Q water, drying at room temperature and heating at 700 °C for several minutes in an oven. Quartz wool was used as a stopper to fix the position of the impregnated AC within the quartz tube.

2.3. Loading of Mercury on Activated Carbon Traps

Mercury was loaded on the impregnated AC traps and analyzed after thermal decomposition/combustion using AAS determination. Elemental mercury (Hg^0) was produced by the reduction of a known amount of the reference material NIST 3133 using tin(II) chloride solution (2 mL of 10% SnCl$_2$ (w/v; for analysis, Merck, Darmstadt, Germany) in 10% HCl (v/v; for analysis, Merck, Darmstadt, Germany)). Produced elemental mercury was loaded onto the carbon trap (trap A, Scheme 2) by purging for 10 min using an airflow of 4 L min^{-1}. The amount of SnCl$_2$ in 10% HCl (v/v), the volume of this solution and the purging time were previously optimized to achieve quantitative reduction of Hg(II) from the aqueous solution. A high pumping rate (4 L min^{-1}) was used to eliminate the possibility of water vapor condensation in the analytical cell of the AAS detector (Scheme 2) [11].

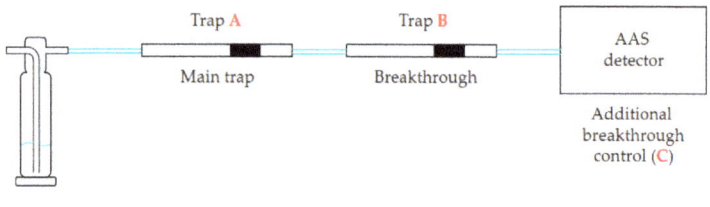

Scheme 2. General scheme of Hg^0 generation via reduction of NIST 3133-traceable Hg(II) standard solution with SnCl$_2$, capturing of produced Hg^0 on activated carbon trap (A), and checking possible breakthrough using additional activated carbon trap (B) and additional breakthrough control (C, AAS detector).

To verify whether mercury had been quantitatively trapped on the trap A, an additional trap (trap B, Scheme 2) was placed on-line after the trap A. Furthermore, an atomic absorbance spectrometer (AAS; Lumex, Lumex Scientific, St. Petersburg, Russia) was occasionally connected to the exit of the trap B and served as an additional breakthrough control (C, Scheme 2). The purging solution was always re-purged to check the purging efficiency (i.e., whether Hg was quantitatively purged out of solution). The amount of mercury present in the purging solution was almost always below detection limit of the AAS detector (at most 0.26% of the original Hg amount).

Using the described approach, 10–2000 ng of Hg^0 was purged onto AC traps. We set the upper limit to 2000 ng because for the determination of Hg in atmospheric samples from pristine and moderately contaminated areas, this amount is great enough for reliable calibration. In addition, this value corresponds to the upper end of the reported working range (2000 ng of Hg on 0.2 g of AC equals to 10,000 ng g^{-1}). The reported working range for solid samples is from 0.5–10,000 ng g^{-1} (when using up to 0.5 g of sample) [11].

2.4. Determination of Mercury on Activated Carbon Traps

The determination of mercury on iodinated AC traps was performed using an AAS detector (Lumex mercury analyzer RA-915M) after thermal desorption at 700 °C in a Lumex PYRO-915+ thermal decomposition attachment (Lumex Scientific, St. Petersburg, Russia) [11]. Instrumental limit of detection (LOD) was 0.3 ng (based on three times the standard deviation of 10 system blanks).

Desorption of Hg from the AC was achieved by heating the sample in a Lumex PYRO-915+ thermal decomposition attachment after the quantitative transfer of the carbon sample and the quartz wool from the trap to an analytical quartz boat. Both iodinated AC traps (A and B; Scheme 2) were measured in the same manner. The detector's response for mercury at different concentration levels was corrected for the signal obtained in the procedural blank, i.e., in the AC trap that was purged with 0 ng of Hg.

To determine the exact concentration or the amount of mercury present on the carbon trap, the analytical system must be properly calibrated. Therefore, we compared the instrument's response for Hg in the carbon traps obtained by the purging of the reduced NIST 3133 standard with the instrument's response for Hg obtained by the decomposition of the directly spiked NIST 3133 standard onto the AC sample. An aliquot (10–30 µL) of the diluted NIST 3133 standard with a known concentration was directly spiked onto the AC sample in the quartz boat, thermally decomposed/combusted at 700 °C using a Lumex PYRO-915+ thermal decomposition attachment and determined using an AAS detector Lumex mercury analyzer RA-915M. The detector's response for mercury at each concentration level was corrected for the signal obtained in the procedural blank, i.e., in the AC trap that was spiked with a Hg-free solution.

The statistical differences between the detector's response for Hg signal in impregnated AC, obtained by purging of Hg^0 and spiking of Hg(II), were examined using paired t-test (SigmaPlot, version 12.0, Systat Software, Erkrath, Germany).

2.5. Determination of Mercury in Real Sample Using Impregnated Activated Carbon Traps—Proof of Concept

To assess the validity of the used approach in the determination of mercury, atmospheric mercury concentration in a real sample was determined using impregnated AC traps (AIC-500). An atmospheric sample with high Hg concentration was used due to the relatively high amount of mercury (minimum 50 ng) that is required for the determination of Hg with a high degree of accuracy and precision using the described method. Therefore, we collected atmospheric samples from a contaminated indoor site containing 68.2 ± 0.95 ng m^{-3} of Hg (average ± standard deviation of individual measurements at 10-min intervals). This concentration was determined separately using the Lumex mercury analyzer RA-915M (continuous mode) that was running during the collection of Hg on AC traps. Atmospheric mercury was loaded onto the impregnated AC trap by purging for 240 min through a soda lime trap (to remove humidity) at an airflow of 3.3 L min^{-1}. The experiment was performed in two replicates. The amount of Hg on AC traps was determined as described above. Accuracy of the method was tested by the determination of Hg in reference material ERM EF412 (brown coal; certified value 70.0 ± 11.0 ng g^{-1}) [31].

2.6. Thermal Fractionation of Mercury on Activated Carbon Traps

Thermal fractionation is a simple method for the estimation of individual mercury compounds released with increasing temperature. A sample of impregnated AC loaded with a known amount of Hg was transferred to a quartz boat and subjected to a heat gradient of 10 °C min^{-1} (from room temperature to 800 °C) under an argon flow of 0.2 L min^{-1}. The released mercury compounds were converted to elemental mercury using a Lumex PYRO-915+ thermal decomposition attachment at 700 °C and detected using an AAS detector (Lumex mercury analyzer RA-915M). The AAS signal was obtained using Rapid software (version 1.00.442, Lumex Scientific, St. Petersburg, Russia) at 1 s resolution and was later converted to fractionation thermogram (AAS signal-temperature relation) by applying heat gradient factor [32].

2.7. Peak Deconvulation of Fractionation Thermograms

To assess how many individual Hg compounds are present in the loaded iodinated AC, peak deconvolution of the fractionation thermograms was applied. First, the obtained fractionation thermogram (at 0.17 °C resolution) was smoothened by applying a locally estimated scatterplot smoothing (LOESS) algorithm using SigmaPlot software. The obtained smoothed scatterplot was used to estimate individual constituent peaks. Peak deconvolution was performed by applying the weighted least squares method using Fityk software (version 1.3.1, M. Wojdyr, Warsaw, Poland) and assuming log-normal distribution of data [33].

3. Results and Discussion

3.1. Comparison of the Detectors Response for Mercury on Activated Carbon Traps Using Two Methods

To estimate whether direct spiking of NIST 3133-treaceable Hg standards on AC traps can be used for the calibration of the system, we compared the detector's response for mercury that was spiked on a trap against the detector's response for mercury purged on carbon traps (sum of Hg amount present on traps A and B; Scheme 2). The results shown in Figure 1 clearly indicate a linear response between these two methods of loading mercury on iodinated AC traps (AIC-500) at different concentration levels. Furthermore, the slope of the linear regression line is 1.0111 with the R^2 value of 0.9986 indicating an almost identical detector's response for both methods. There were no statistically significant differences in the detector's response for the Hg signal in impregnated AC obtained by the purging of Hg^0 and the spiking of $Hg(II)$ ($p = 0.724$; paired t-test).

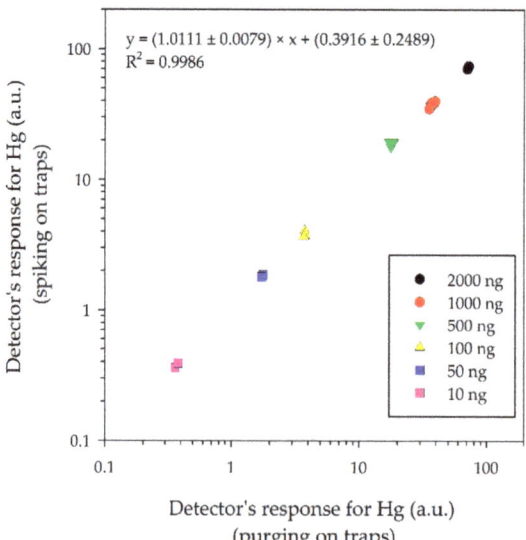

Figure 1. Linear dependence of the detector's signal for Hg on activated carbon traps (AIC-500) obtained by two loading methods (purging onto carbon traps and spiking on activated carbon). a.u.—arbitrary units.

These results indicate that mercury that is purged from the atmospheric sample on the AC trap can be easily determined using combustion, while the analytical system can be calibrated by the simple spiking of the AC using the appropriate dilution of the reference standard material NIST 3133. Direct spiking of NIST 3133-treaceable mercury standard solution on carbon can, therefore, be used for the traceable calibration of analytical systems for the determination of mercury purged on carbon traps.

3.2. Comparison of Hg Adsorption on Different Impregnated Activated Carbon Traps

To test the robustness and versatility of the method, we tested different iodinated AC traps. The results from the adsorption and breakthrough on AIC-500 carbon traps are presented alongside the results from other iodinated AC traps to facilitate data comparison (Table 1). To test whether other impregnated AC traps behave in the similar manner, we prepared several in-house impregnated AC traps. The detector's response for mercury purged to different impregnated AC traps agrees with the signal obtained by the direct spiking of the standard reference material at 1000 ng level. We determined which traps and corresponding impregnating solutions are the best for achieving quantitative mercury trapping and which impregnated AC traps show the lowest breakthrough of mercury at the 1000 ng level. The comparison of the results for different impregnated AC traps is presented in Table 1.

Table 1. Performance of commercially available and in-house prepared activated carbon traps. All results refer to 1000 ng of purged/spiked Hg.

Activated Carbon/Impregnating Solution	Trap A (%) [1]	Trap B (%) [1]
AIC-500 (n = 9)	100 ± 1.78	<LOD [2] – 0.10
0.10 mol L^{-1} NH$_4$I (n = 7)	75.7 ± 6.32	0.71 ± 0.22
0.03 mol L^{-1} NH$_4$I (n = 14)	101 ± 3.25	0.74 ± 0.89
0.01 mol L^{-1} NH$_4$I (n = 15)	99.3 ± 2.52	<LOD [2] – 0.27

[1] Relative to spiking with Hg(II); Trap A and Trap B refer to activated carbon traps, as described in Scheme 2; [2] limit of detection (LOD) was between 0.5 and 2.1 ng, depending on Hg impurities in carbon.

As seen in Table 1, commercially available AC (AIC-500) and in-house prepared AC impregnated with 0.01 mol L^{-1} NH$_4$I solution show the best performance. Both traps show quantitative capturing of mercury at 1000 ng level and exceptionally low breakthrough, as seen by the amount of mercury adsorbed on trap B. It is important to note that the additional breakthrough control (C) was always within the noise of the AAS detector. In addition, the amount of mercury present in the purging solution was always up to 0.26% of the original Hg amount present in the purging vessel (impinger) indicating that mercury was completely purged out onto the AC trap. Due to the quantitative recovery of the Hg purged to the carbon traps relative to spiking, we presume that spiking can be used as a reliable calibration method for the determination of Hg on impregnated AC traps.

We decided to use only AIC-500 traps for the determination of mercury in the real atmospheric sample and for the assessment of linearity (Figure 1) due to slightly better results compared to in-house prepared AC traps (better repeatability on trap A and the lowest breakthrough on trap B (Table 1)). We presume that the calibration is also linear up to 1000 ng when using in-house prepared AC treated with 0.01 mol L^{-1} NH$_4$I and 0.03 mol L^{-1} NH$_4$I. This assumption is based on the fact that AAS signals for AIC-500 and mentioned AC traps are similar at the 1000 ng level (within ±3.33% of the average signal for AIC-500). However, these linearities were not tested, as explained above.

High amounts of iodine seem to affect Hg determinations, as seen by low Hg recoveries on AC treated with 0.10 mol L^{-1} (Table 1). This might be due to interferences between mercury and iodine absorption lines (254 and 256 nm, respectively) [34]. In cases when complex matrixes are analyzed, strong background absorption arises due to the production of large amounts of smoke and interference radicals. The observed strong background absorption cannot be corrected using the inherent selectivity of the Zeeman atomic absorption spectrometer [11]. Low iodine amounts show quantitative trapping and do not considerably affect Hg determinations by the AAS detector. The amount of iodine bound on AC treated with impregnating solution with the lowest iodine content (0.01 mol L^{-1} NH$_4$I) was estimated from the difference between the amount of iodine in this solution prior to and after the equilibration of AC. The concentration of iodine in the impregnating solution was lowered from 0.01 mol L^{-1} prior to equilibration to 0.0078 mol L^{-1} after equilibration. This change corresponds to the adsorption of 0.21 mmol of iodide per gram of AC (i.e., mass fraction of iodide 2.69%). Comparison of this value with the literature is rather difficult: commercial AC traps do not disclose the amount of the bound halogen of activated carbon, while scientific articles usually do not report the content. The most comprehensive review of sorbents for mercury removal from flue gas [16] reported iodinated AC that contained 3.5% of iodine, which is slightly higher than the iodine content in in-house prepared iodinated AC. Higher iodine content on AC might be useful for the removal of mercury from flue gas, but it is not appropriate for the determination of low atmospheric Hg concentrations, as described above.

Even though iodine in iodinated AC might cause problems during Hg determinations, it also creates much more stable Hg complexes compared to other halogens. Therefore, it is important to balance between analytical performance and the stability of Hg complexes when using iodinated AC traps. AC treated with impregnating solution containing the lowest iodine content (0.01 mol L^{-1} NH$_4$I) seems to be right at the balance, as demonstrated by good analytical performance at 1000 ng level (Table 1). Even though this balancing would not be required when using chlorinated or brominated ACs, we intentionally wanted to test this approach using NH$_4$I due to the specific composition of the reaction product (Millon's iodide).

Gaseous Hg(II) can be captured on AC traps [20], but it cannot be separately determined as the whole AC from the trap is quantitatively transferred to an analytical quartz boat and subjected to combustion. Hg(II) can be only determined using speciation traps (solid KCl + AC). In addition, the loading of traceable amounts of gaseous Hg(II) at low atmospheric concentrations is still under development [21].

3.3. Determination of Mercury in Real Atmospheric Sample—Proof of Concept

An atmospheric sample from contaminated indoor site containing 68.2 ± 0.95 ng m^{-3} of Hg was collected on the AIC-500 activated carbon trap. Based on this concentration, purging time and airflow through the trap, it was calculated that 54.0 ± 0.99 ng of Hg was collected on the AC trap (average ± standard deviation of two parallel measurements). The response of the AAS detector was compared with the corresponding response for the NIST 3133 spike onto AC (50 ng of Hg). Following the subtraction of corresponding blanks, the AC trap contained 53.2 ± 3.05 ng of Hg. The accuracy of Hg determinations was tested using reference material ERM EF412 (brown coal). The obtained value (69.7 ± 2.86 ng g^{-1}; n = 4) was within the uncertainty of its certified value (70.0 ± 11.0 ng g^{-1}).

Although the obtained average value of Hg content in the real atmospheric sample is slightly lower than the calculated amount of Hg, it is still within the standard deviation of measurements. The possible reason for this observation is Hg adsorption on soda lime. Although normally this never represents an issue, long sampling time (240 min) might have caused the adsorption of a significant amount of water vapor from the air, and possibly some Hg might have been dissolved in this water. This is especially important for gaseous oxidized mercury, which is readily soluble in water [15]. The relatively high standard deviation of the determined Hg on the AC trap can be attributed to the variability in the spiking volume (10 µL), which should be as smallest as possible.

The determination of Hg using AC traps can be readily used for atmospheric samples containing elevated Hg concentrations with a high degree of accuracy and precision using the described method. However, for samples with ambient atmospheric concentrations (about 1.5 ng m^{-3}) [35–38], longer purging time or higher airflows are required. This might cause unwanted Hg transformations during sample collection or the breakthrough of mercury (especially in case of high airflow). These issues can be overcome by sampling lower amounts of mercury and/or using a more sensitive detector (e.g., atomic fluorescence spectrometer). However, the level cannot be extremely low, due to the relatively high blanks of impregnated AC (Table 1).

In principle, this method could be applied for the characterization of Hg transformations in atmospheric samples using the stable isotope fractionation approach. The method is limited by the relatively high LOD; for reliable determination of stable isotope ratios, blanks should ideally account for less than 2% of the collected amount of atmospheric Hg [39,40]. Furthermore, desorption of mercury from the AC traps must be quantitative to avoid Hg fractionation on the AC. This is readily achieved using combustion at a high temperature. Released Hg must be quantitatively trapped in a solution that can completely oxidize Hg^0 to $Hg(II)$ (e.g., solution of potassium permanganate or mixture of nitric and hydrochloric acid) [12]. Exit from the Lumex PYRO-915+ thermal decomposition attachment could be connected to this trapping solution to quantitatively collect combustion products. As the calibration is performed immediately after the determination of Hg in samples, a combusted NIST 3133 standard could be trapped in this solution and used directly as a bracketing standard.

3.4. Thermal Fractionation of Hg on Impregnated Activated Carbon

Impregnation of AC using NH_4I solutions and not with commonly used KI solutions was performed because the former readily reacts with Hg and forms the insoluble iodide of the Millon's base, $[Hg_2N]I \cdot H_2O$. Its literature decomposition temperature of 350–400 °C is close to the decomposition temperature (about 370 °C) obtained using thermal fractionation [41]. The obtained thermograms for AIC-500 and in-house prepared iodinated AC (0.01 mol L^{-1} NH_4I) are presented in Figure 2. In-house prepared iodinated AC has a higher maximum temperature for the highest peak. However, it also starts to release mercury at a lower temperature (about 150 °C) compared to AIC-500.

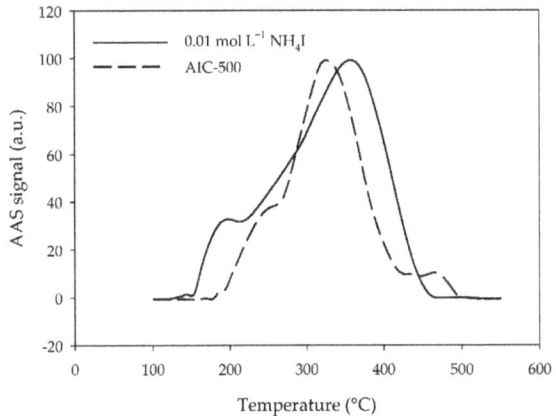

Figure 2. Fractionation thermograms for AIC-500 and in-house prepared iodinated activated carbon (0.01 mol L^{-1} NH_4I). Parts of the thermogram below 100 °C and above 550 °C were removed for clarity as the values for AAS signal were about zero. a.u.—arbitrary units.

To assess how many peaks (i.e., Hg compounds) are present under each curve in Figure 2, peak deconvolution was performed using the weighted least squares method [33]. Assuming log-normal distribution of the constituent peaks, we identified that each thermogram is comprised of three individual peaks (Figure 3). However, without direct comparison with pure Hg standards on impregnated AC, it is not possible to exactly determine the exact composition of the three individual mercury compounds (not in the scope of this work).

Based on the literature data, we assume that the Hg bound to in-house impregnated iodinated AC is in form of Millon's base due to the similar theoretical decomposition temperature. The iodide of Millon's base decomposes on heating; decomposition starts at a temperature of 160 °C [41], which is the temperature at which the first peak starts to appear during the thermal fractionation of in-house prepared AC (Figure 3b). The second deconvoluted peak at 280 °C might be attributed to the decomposition of an unidentified intermediate degradation compound, as the structure of Millon's base is polymeric in nature and the chemical formula $[Hg_2N]^+$ does not represent an exact ionic species. Millon's base is composed of a silica-like network of N and Hg a in four- and two-coordination, respectively, with anion and water (if present) in the interstitial spaces [42]. Furthermore, the decomposition is not rapid until about 350–400 °C [41], where we have observed the third peak. Slow decomposition towards the maximum is one of the reasons why we applied the log-normal distribution of the constituent peaks during the peak deconvolution analysis.

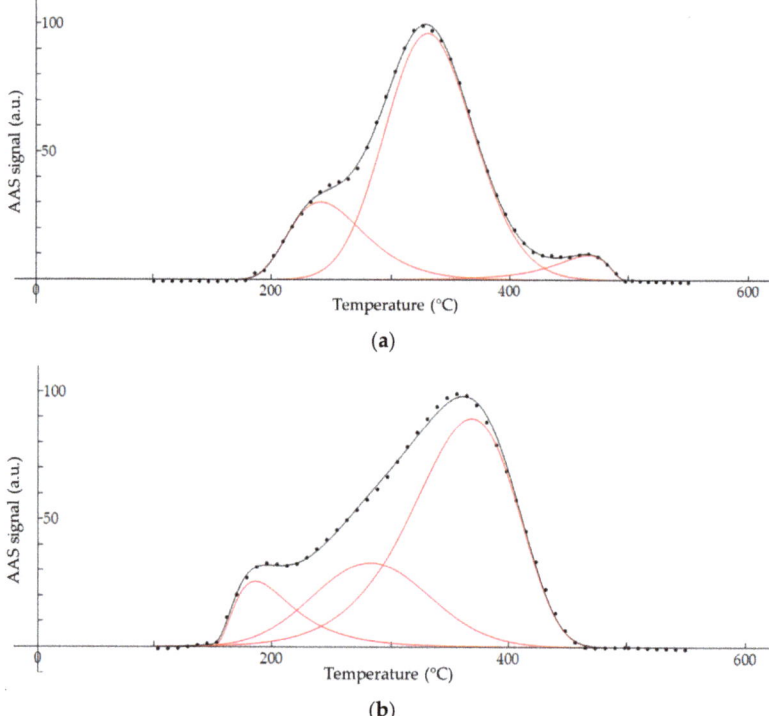

Figure 3. Peak deconvolution of fractionation thermograms for (**a**) AIC-500 and (**b**) in-house prepared iodinated activated carbon (0.01 mol L^{-1} NH_4I) assuming log-normal distribution of constituent peaks. Black dots represent smoothed experimental points (weighted least squares method), red curves deconvoluted peaks and the black curve the summarized model (sum of red curves). Parts of the thermograms below 100 °C and above 550 °C were removed for clarity as the values for AAS signal were about zero. a.u.—arbitrary units.

Most binary transition-metal–nitrogen compounds are highly endothermic compounds, including Hg–N compounds [43]. The possible explosive decomposition of these nitrogen-rich compounds is due to their extremely low energy barriers [44]. The decomposition of the Millon's iodide is similar to the thermal decomposition of other salts of Millon's base. $[Hg_2N]NO_3$ has a similar decomposition temperature (380 °C) [45], while $[Hg_2N]N_3$ exhibits a smooth decomposition accompanied by the release of molecular nitrogen and Hg^0 [43]. Nevertheless, the decomposition temperature of $[Hg_2N]I$ on activated carbon should be taken with a high degree of reservation, because it was recently shown that matrix effect greatly influences the decomposition of various Hg compounds by shifting the temperature of Hg released from the sample [32].

Although Hg is purged onto iodinated AC traps as Hg^0, it is most probably oxidized to Hg(II) on the surface. The dynamic adsorption of mercury on iodinated AC traps suggests complex mechanisms for the adsorption [46]. Physical adsorption is probably the first step in the removal of mercury by both virgin and impregnated ACs [16]. Recently, it was demonstrated that purged Hg^0 is readily oxidized to Hg(II) on the brominated AC surface at 30 °C and 140 °C, indicating that chemisorption is the likely adsorption mechanism of Hg [47]. As Hg^0 removal efficiency increases with adsorption temperature, it is assumed that Hg^0 vapors chemically react with impregnated iodine on the AC surface (rates of chemisorption increase with the rise in temperature) [17]. Desorption of mercury from spent ACs, impregnated with both iodine and KI, suggests that Hg^0 is oxidized by elemental iodine and trapped in the form of K_2HgI_4 and $KHgI_3$ [48]. We presume that iodine also acts as an oxidant in in-house impregnated iodinated AC, as iodide has a low standard reduction potential (+0.54 V) and might be oxidized to elemental iodine by oxygen from the moist air. Consequently, Hg^0 is probably oxidized to HgI_4^{2-} and HgI_3^-, but in contrast to KI, the presence of ammonium ion converts these Hg-iodide complexes to Millon's iodide, $[Hg_2N]I$. Gaseous Hg(II) can also be trapped on iodinated AC and is probably first converted to HgI_2, then to Hg-iodide complexes, and finally to $[Hg_2N]I$. The formation of HgI_2 enables impregnated AC to remove even more elemental mercury [16]. We presume that GOM and GEM are readily converted to Hg-iodide complexes on the AC surface and later fixed in the form of Millon's iodide, thus providing additional active sites for further adsorption of Hg on activated carbon.

4. Conclusions

AC can be used for quantitative pre-collection of atmospheric mercury. The determination of mercury bound to AC can be easily achieved using combustion coupled with atomic absorption spectrometry. Comparison of different ACs demonstrated that commercial iodinated AC and activated carbon treated with 0.01 mol L^{-1} NH$_4$I show quantitative capturing of Hg^0 (99.3–100%) while having the lowest breakthrough (<LOD – 0.27%). AC containing a greater amount of iodine shows lower Hg recoveries, probably due to interferences during analysis using atomic absorption. The identical response of the AAS detector for two Hg loading methods (purging of Hg^0 and spiking SRM solution) in a wide concentration range (10–2000 ng) indicated that the spiking of SRM on AC can be readily used for the calibration of the analytical system used for the determination of atmospheric mercury. The concentration of Hg in a real atmospheric sample collected on an iodinated AC trap was determined using spiking calibration and was traceable to SRM NIST 3133. The accuracy of the method was confirmed by a good agreement between the measured Hg content in a reference material ERM EF412 and its certified value. The use of ammonium iodide in impregnating solutions, probably converted Hg on AC to the iodide of Millon's base, as the positions of deconvoluted peaks from fractionation thermogram agreed well with the literature values.

Author Contributions: Conceptualization, I.Ž. and M.H.; methodology, I.Ž. and M.J.; validation, I.Ž., S.B., M.J. and J.K.; formal analysis, I.Ž., M.J. and S.B.; data curation, I.Ž. and J.K.; writing—original draft preparation, I.Ž.; writing—review and editing, S.B., J.K. and M.H.; visualization, I.Ž.; supervision, M.H.; project administration, M.H.; funding acquisition, M.H. All authors have read and agreed to the published version of the manuscript.

Funding: This research was funded by EMPIR, grant number 16ENV01 (MercOx); Slovenian Research Agency, grant number P1-0143.

Acknowledgments: The authors would like to thank Warren T. Corns for donation in kind of iodinated activated carbon.

Conflicts of Interest: The authors declare no conflict of interest. The funders had no role in the design of the study; in the collection, analyses, or interpretation of data; in the writing of the manuscript, or in the decision to publish the results.

References

1. Park, J.-D.; Zheng, W. Human Exposure and Health Effects of Inorganic and Elemental Mercury. *J. Prev. Med. Public Health* **2012**, *45*, 344–352. [CrossRef] [PubMed]

2. Fernandes Azevedo, B.; Barros Furieri, L.; Peçanha, F.M.; Wiggers, G.A.; Frizera Vassallo, P.; Ronacher Simões, M.; Fiorim, J.; Rossi de Batista, P.; Fioresi, M.; Rossoni, L.; et al. Toxic Effects of Mercury on the Cardiovascular and Central Nervous Systems. *J. Biomed. Biotechnol.* **2012**, *2012*, 1–11. [CrossRef] [PubMed]
3. Amos, H.M.; Jacob, D.J.; Streets, D.G.; Sunderland, E.M. Legacy impacts of all-time anthropogenic emissions on the global mercury cycle. *Biogeochem. Cycles* **2013**, *27*, 410–421. [CrossRef]
4. Mason, R.P.; Choi, A.L.; Fitzgerald, W.F.; Hammerschmidt, C.R.; Lamborg, C.H.; Soerensen, A.L.; Sunderland, E.M. Mercury biogeochemical cycling in the ocean and policy implications. *Environ. Res.* **2012**, *119*, 101–117. [CrossRef] [PubMed]
5. Ariya, P.A.; Amyot, M.; Dastoor, A.; Deeds, D.; Feinberg, A.; Kos, G.; Poulain, A.; Ryjkov, A.; Semeniuk, K.; Subir, M.; et al. Mercury Physicochemical and Biogeochemical Transformation in the Atmosphere and at Atmospheric Interfaces: A Review and Future Directions. *Chem. Rev.* **2015**, *115*, 3760–3802. [CrossRef] [PubMed]
6. Gustin, M.S.; Evers, D.C.; Bank, M.S.; Hammerschmidt, C.R.; Pierce, A.; Basu, N.; Blum, J.; Bustamante, P.; Chen, C.; Driscoll, C.T.; et al. Importance of Integration and Implementation of Emerging and Future Mercury Research into the Minamata Convention. *Environ. Sci. Technol.* **2016**, *50*, 2767–2770. [CrossRef]
7. Obrist, D.; Kirk, J.L.; Zhang, L.; Sunderland, E.M.; Jiskra, M.; Selin, N.E. A review of global environmental mercury processes in response to human and natural perturbations: Changes of emissions, climate, and land use. *Ambio* **2018**, *47*, 116–140. [CrossRef]
8. Bank, M.S. *Mercury in the Environment: Pattern and Process*; University of California Press: Los Angeles, CA, USA, 2012; ISBN 9780520271630.
9. Ren, X.; Luke, W.T.; Kelley, P.; Cohen, M.D.; Olson, M.L.; Walker, J.; Cole, R.; Archer, M.; Artz, R.; Stein, A.A. Long-Term Observations of Atmospheric Speciated Mercury at a Coastal Site in the Northern Gulf of Mexico during 2007–2018. *Atmosphere* **2020**, *11*, 268. [CrossRef]
10. Sprovieri, F.; Hedgecock, I.M.; Pirrone, N. An investigation of the origins of reactive gaseous mercury in the Mediterranean marine boundary layer. *Atmos. Chem. Phys.* **2010**, *10*, 3985–3997. [CrossRef]
11. Sholupov, S.; Pogarev, S.; Ryzhov, V.; Mashyanov, N.; Stroganov, A. Zeeman atomic absorption spectrometer RA-915+ for direct determination of mercury in air and complex matrix samples. *Fuel Process. Technol.* **2004**, *85*, 473–485. [CrossRef]
12. Fu, X.; Heimbürger, L.E.; Sonke, J.E. Collection of atmospheric gaseous mercury for stable isotope analysis using iodine-and chlorine-impregnated activated carbon traps. *J. Anal. At. Spectrom.* **2014**, *29*, 841–852. [CrossRef]
13. Bérail, S.; Cavalheiro, J.; Tessier, E.; Barre, J.P.G.; Pedrero, Z.; Donard, O.F.X.; Amouroux, D. Determination of total Hg isotopic composition at ultra-trace levels by on line cold vapor generation and dual gold-amalgamation coupled to MC-ICP-MS. *J. Anal. At. Spectrom.* **2017**, *32*, 373–384. [CrossRef]
14. Horvat, M. Determination of Mercury and its Compounds in Water, Sediment, Soil and Biological Samples. In *Dynamics of Mercury Pollution on Regional and Global Scales*; Springer: New York, NY, USA, 2005; pp. 153–190.
15. Huang, J.; Lyman, S.N.; Hartman, J.S.; Gustin, M.S. A review of passive sampling systems for ambient air mercury measurements. *Environ. Sci. Process. Impacts* **2014**, *16*, 374–392. [CrossRef] [PubMed]
16. Granite, E.J.; Pennline, H.W.; Hargis, R.A. Novel Sorbents for Mercury Removal from Flue Gas. *Ind. Eng. Chem. Res.* **2000**, *39*, 1020–1029. [CrossRef]
17. Lee, S.J.; Seo, Y.-C.; Jurng, J.; Lee, T.G. Removal of gas-phase elemental mercury by iodine- and chlorine-impregnated activated carbons. *Atmos. Environ.* **2004**, *38*, 4887–4893. [CrossRef]
18. Musmarra, D.; Karatza, D.; Lancia, A.; Prisciandaro, M.; Di Celso, G.M. A comparison among different sorbents for mercury adsorption from flue gas. *Chem. Eng. Trans.* **2015**, *43*, 2461–2466. [CrossRef]
19. Yu, J.-G.; Yue, B.-Y.; Wu, X.-W.; Liu, Q.; Jiao, F.-P.; Jiang, X.-Y.; Chen, X.-Q. Removal of mercury by adsorption: A review. *Environ. Sci. Pollut. Res.* **2016**, *23*, 5056–5076. [CrossRef]
20. US EPA. Method 30B—Mercury Sorbent Trap Procedure. Available online: https://www.epa.gov/sites/production/files/2017-08/documents/method_30b.pdf (accessed on 3 May 2018).
21. Saxholm, S.; Rajamäki, T.; Hämäläinen, J.; Hildén, P. Dynamic calibration method for reactive gases. *Meas. Sci. Technol.* **2019**, *31*, 1–14. [CrossRef]
22. Dumarey, R.; Temmerman, E.; Adams, R.; Hoste, J. The accuracy of the vapour-injection calibration method for the determination of mercury by amalgamation/cold-vapour atomic absorption spectrometry. *Anal. Chim. Acta* **1985**, *170*, 337–340. [CrossRef]

23. Dumarey, R.; Brown, R.J.C.; Corns, W.T.; Brown, A.S.; Stockwell, P.B. Elemental mercury vapour in air: The origins and validation of the 'Dumarey equation' describing the mass concentration at saturation. *Accredit. Qual. Assur.* **2010**, *15*, 409–414. [CrossRef]
24. Huber, M.L.; Laesecke, A.; Friend, D.G. Correlation for the Vapor Pressure of Mercury. *Ind. Eng. Chem. Res.* **2006**, *45*, 7351–7361. [CrossRef]
25. Gustin, M.S.; Huang, J.; Miller, M.B.; Peterson, C.; Jaffe, D.A.; Ambrose, J.; Finley, B.D.; Lyman, S.N.; Call, K.; Talbot, R.; et al. Do we understand what the mercury speciation instruments are actually measuring? Results of RAMIX. *Environ. Sci. Technol.* **2013**, *47*, 7295–7306. [CrossRef] [PubMed]
26. Quétel, C.R.; Zampella, M.; Brown, R.J.C. Temperature dependence of Hg vapour mass concentration at saturation in air: New SI traceable results between 15 and 30°C. *TrAC—Trends Anal. Chem.* **2016**, *85*, 81–88. [CrossRef]
27. US EPA. Method 1631, Revision E: Mercury in Water by Oxidation, Purge and Trap, and Cold Vapor Atomic Fluorescence Spectrometry. Available online: https://www.epa.gov/sites/production/files/2015-08/documents/method_1631e_2002.pdf (accessed on 2 February 2020).
28. Berisha, S.; Živković, I.; Kotnik, J.; Mlakar, T.L.; Horvat, M. Quantification of total mercury in samples from cement production processing with thermal decomposition coupled with AAS. *Accredit. Qual. Assur.* **2020**, *25*, 233–242. [CrossRef]
29. Takaku, Y.; Shimamura, T.; Masuda, K.; Igarashi, Y. Iodine Determination in Natural and Tap Water Using Inductively Coupled Plasma Mass Spectrometry. *Anal. Sci.* **1995**, *11*, 823–827. [CrossRef]
30. Jerše, A.; Jaćimović, R.; Maršić, N.K.; Germ, M.; Širclej, H.; Stibilj, V. Determination of iodine in plants by ICP-MS after alkaline microwave extraction. *Microchem. J.* **2018**, *137*, 355–362. [CrossRef]
31. ERM-EF412 Brown Coal (GCV, Ash, Volatile Matter, Elements). Available online: https://crm.jrc.ec.europa.eu/p/40455/40467/By-material-matrix/Fuels/ERM-EF412-BROWN-COAL-GCV-Ash-Volatile-Matter-Elements/ERM-EF412 (accessed on 1 May 2020).
32. Sedlar, M.; Pavlin, M.; Popovič, A.; Horvat, M. Temperature stability of mercury compounds in solid substrates. *Open Chem.* **2014**, *13*, 404–419. [CrossRef]
33. Wojdyr, M. Fityk: A general-purpose peak fitting program. *J. Appl. Crystallogr.* **2010**, *43*, 1126–1128. [CrossRef]
34. Barnett, N.W.; Kirkbright, G.F. Electrothermal vaporisation sample introduction into an atmospheric pressure helium microwave-induced plasma for the determination of iodine in hydrochloric acid. *J. Anal. At. Spectrom.* **1986**, *1*, 337. [CrossRef]
35. Cizdziel, J.; Jiang, Y.; Nallamothu, D.; Brewer, J.; Gao, Z. Air/Surface Exchange of Gaseous Elemental Mercury at Different Landscapes in Mississippi, USA. *Atmosphere* **2019**, *10*, 538. [CrossRef]
36. Gratz, L.; Eckley, C.; Schwantes, S.; Mattson, E. Ambient Mercury Observations near a Coal-Fired Power Plant in a Western U.S. Urban Area. *Atmosphere* **2019**, *10*, 176. [CrossRef] [PubMed]
37. Slemr, F.; Brunke, E.-G.; Ebinghaus, R.; Temme, C.; Munthe, J.; Wängberg, I.; Schroeder, W.; Steffen, A.; Berg, T. Worldwide trend of atmospheric mercury since 1977. *Geophys. Res. Lett.* **2003**, *30*, 1–4. [CrossRef]
38. Sprovieri, F.; Pirrone, N.; Ebinghaus, R.; Kock, H.; Dommergue, A. A review of worldwide atmospheric mercury measurements. *Atmos. Chem. Phys.* **2010**, *10*, 8245–8265. [CrossRef]
39. Liu, H.; Yu, B.; Yang, L.; Wang, L.; Fu, J.; Liang, Y.; Bu, D.; Yin, Y.; Hu, L.; Shi, J.; et al. Terrestrial mercury transformation in the Tibetan Plateau: New evidence from stable isotopes in upland buzzards. *J. Hazard. Mater.* **2020**, *400*, 1–6. [CrossRef] [PubMed]
40. Huang, Q.; Reinfelder, J.R.; Fu, P.; Huang, W. Variation in the mercury concentration and stable isotope composition of atmospheric total suspended particles in Beijing, China. *J. Hazard. Mater.* **2020**, *383*, 1–7. [CrossRef]
41. Weiser, H.B. The Luminescence of the Iodide of Millon's Base. *J. Phys. Chem.* **1917**, *21*, 37–47. [CrossRef]
42. Urben, P.G. *Bretherick's Handbook of Reactive Chemical Hazards*, 8th ed.; Elsevier: Amsterdam, The Netherlands, 2017.
43. Lund, H.; Oeckler, O.; Schröder, T.; Schulz, A.; Villinger, A. Mercury Azides and the Azide of Millon's Base. *Angew. Chemie Int. Ed.* **2013**, *52*, 10900–10904. [CrossRef]
44. Tornieporth-Oetting, I.C.; Klapötke, T.M. Covalent Inorganic Azides. *Angew. Chemie Int. Ed.* **1995**, *34*, 511–520. [CrossRef]

45. Nockemann, P.; Meyer, G. Formation of $NH_4[Hg_3(NH)_2](NO_3)_3$ and Transformation to $[Hg_2N](NO_3)$. *Z. Anorg. Allg. Chem.* **2002**, *628*, 2709–2714. [CrossRef]
46. Matsumura, Y. Adsorption of mercury vapor on the surface of activated carbons modified by oxidation or iodization. *Atmos. Environ.* **1974**, *8*, 1321–1327. [CrossRef]
47. Sasmaz, E.; Kirchofer, A.; Jew, A.D.; Saha, A.; Abram, D.; Jaramillo, T.F.; Wilcox, J. Mercury chemistry on brominated activated carbon. *Fuel* **2012**, *99*, 188–196. [CrossRef]
48. Granite, E.J.; Pennline, H.W.; Hargis, R.A. *Sorbents for Mercury Removal from Flue Gas*; U.S. Department of Energy: Pittsburgh, PA, USA, 1998.

© 2020 by the authors. Licensee MDPI, Basel, Switzerland. This article is an open access article distributed under the terms and conditions of the Creative Commons Attribution (CC BY) license (http://creativecommons.org/licenses/by/4.0/).

Article

Gaseous Elemental Mercury Concentrations along the Northern Gulf of Mexico Using Passive Air Sampling, with a Comparison to Active Sampling

Byunggwon Jeon [1], James V. Cizdziel [1,*], J. Stephen Brewer [2], Winston T. Luke [3], Mark D. Cohen [3], Xinrong Ren [3,4] and Paul Kelley [3,4]

1. Department of Chemistry and Biochemistry, University of Mississippi, University, MS 38677, USA; bjeon@go.olemiss.edu
2. Department of Biology, University of Mississippi, University, MS 38677, USA; jbrewer@olemiss.edu
3. Air Resources Laboratory, National Oceanic and Atmospheric Administration, College Park, MD 20740, USA; winston.luke@noaa.gov (W.T.L.); mark.cohen@noaa.gov (M.D.C.); xinrong.ren@noaa.gov (X.R.); paul.kelley@noaa.gov (P.K.)
4. Department of Atmospheric and Oceanic Science, University of Maryland, College Park, MD 20742, USA
* Correspondence: cizdziel@olemiss.edu

Received: 4 September 2020; Accepted: 23 September 2020; Published: 26 September 2020

Abstract: Mercury is a toxic element that is dispersed globally through the atmosphere. Accurately measuring airborne mercury concentrations aids understanding of the pollutant's sources, distribution, cycling, and trends. We deployed MerPAS® passive air samplers (PAS) for ~4 weeks during each season, from spring 2019 to winter 2020, to determine gaseous elemental mercury (GEM) levels at six locations along the northern Gulf of Mexico, where the pollutant is of particular concern due to high mercury wet deposition rates and high concentrations in local seafood. The objective was to (1) evaluate spatial and seasonal trends along the Mississippi and Alabama coast, and (2) compare active and passive sampling methods for GEM at Grand Bay National Estuarine Research Reserve, an Atmospheric Mercury Network site. We observed higher GEM levels ($p < 0.05$) in the winter (1.53 ± 0.03 ng m^{-3}) compared to other seasons at all sites; with the general pattern being: winter > spring > summer ≈ fall. Average GEM levels (all deployment combined) were highest at Bay St. Louis (1.36 ± 0.05 ng m^{-3}), the western-most site nearest the New Orleans metropolitan area, and lowest at Cedar Point (1.07 ± 0.09 ng m^{-3}), a coastal marsh with extensive vegetation that can uptake GEM. The MerPAS units compared reasonably well with the established active monitoring system, but gave slightly lower concentrations, except in the winter when the two methods were statistically similar. Both the passive and active sampling methods showed the same seasonal trends and the difference between them for each season was <15%, acceptable for evaluating larger spatial and temporal trends. Overall, this work demonstrates that PASs can provide insight into GEM levels and the factors affecting them along coastal regions.

Keywords: atmospheric mercury; gaseous elemental mercury; passive air sampler; MerPAS®; seasonal trend; spatial trend; Gulf of Mexico; Grand Bay

1. Introduction

Mercury (Hg) is a persistent and toxic pollutant with a complex biogeochemical cycle, where the atmosphere plays an important role, including transport of the contaminant on local, regional, and global scales [1]. The understanding of atmospheric Hg has greatly advanced with the capability to measure gaseous elemental mercury (GEM), gaseous oxidized mercury (GOM), and particle bound mercury (PBM), the three main classes of airborne Hg species. There are challenges in accurately measuring these species and properly interpreting the results [2–4]. GEM, the predominant form typically encompassing >95% of the total gaseous Hg, has a relatively long residence time (~6 months or more) compared to GOM and PBM (~days to weeks) [5]. Levels of each Hg species vary depending on proximity to sources, meteorological conditions, season, and other factors, with GOM and PBM levels plummeting when they are scavenged by precipitation [6,7]. GEM concentrations tend to be more stable, with background levels in the northern hemisphere about 1.5 ng m^{-3} [8]. GEM levels are decreasing at many sites in North America and Europe, likely due to the phase-out of Hg from commercial products, and increased adoption of air pollution control technologies [9]. GEM is slowly converted to PBM and highly soluble and particle-reactive GOM by photochemical and other reactions [5,6]. GOM and PBM concentrations tend to be highest near anthropogenic point sources, especially combustion sources such as coal fired power plants or waste incinerators [2,5,7]. Once deposited to terrestrial and aquatic ecosystems, Hg can be re-emitted or, given the right biogeochemical conditions, converted by certain microorganisms to methyl-Hg, a neurotoxic form that can readily accumulate in organisms and concentrate up the food chain to levels that can harm both wildlife and humans [1,6,10].

With abundant coastal wetlands that promote production and transfer of methyl-Hg into primary producers, the northern Gulf of Mexico (nGoM, a portion of the U.S. Gulf Coast extending from the Suwannee River, in the Florida panhandle, to the Sabine River, near the state line between Louisiana and Texas) is prone to Hg contaminated food webs [11]. Another factor contributing to high Hg levels along the nGoM is that the region consistently has some of the highest wet Hg deposition rates in the USA [12,13]. So, it is not surprising that levels of methyl-Hg in seafood along the nGoM exceed other U.S. coastlines, and that there are widespread fish consumption advisories in the region. This is concerning because (1) nGoM residents tend to consume more seafood than other U.S. residents, with as much as 30% of the coastal population estimated to exceed the U.S. Environmental Protection Agency's reference dose for MeHg [14], and (2) the economy of the region is intricately linked to commercial and recreational fishing. Moreover, we hypothesize that the GoM "dead-zone", a low oxygen area in the waters of the nGoM near the mouth of the Mississippi River and its spillways that occurs each summer as a result of nutrient pollution from agriculture and developed land runoff in the Mississippi River watershed, may exacerbate the Hg problem by producing conditions that favor the production of MeHg, because organic matter and low oxygen fuel certain methylating-microbes [15]. The periodic nature of the dead-zone (oxic-anoxic changes) may affect the speciation and bioavailability of Hg, which, in turn, may affect the net surface exchange of GEM with the atmosphere. Thus, it is important to measure atmospheric Hg at locations along the nGoM to help understand Hg sources, distribution, trends, and cycling in that region.

There is a relatively long record of atmospheric Hg measurements at Grand Bay National Estuarine Research Reserve (NERR) located on the eastern portion of the Mississippi coastline [16]. In addition to long-term speciated-mercury measurements, collected with an automated instrument from Tekran Instruments Corporation (hereafter just Tekran) atop a 10 m tower, along with trace gas and meteorological monitoring, research at the site has included intensive studies on atmospheric mercury speciation [17] and Hg isotopic analyses [18]. The data have provided a valuable insight on atmospheric Hg at the site, including impacts from both local and regional sources as well as large-scale Hg cycling phenomena, species-specific isotopic compositions, and diurnal and seasonal variation in Hg species. As the instrument uses active sampling, the data are temporally rich, allowing correlation with other atmospheric constituents, such as ozone and sulfur dioxide [19]. GEM depletion events have been observed in the early morning at the site, likely due to uptake by plants, and a slight GEM elevation during the day has generally been observed, likely due to downward mixing form higher concentrations aloft [16]. However, the research has been unable to directly address spatial variability in GEM concentrations because the Tekran instrument is stationary, costly, and requires power.

Passive air sampling is a low-cost no-power alternative approach to active sampling. In passive sampling the gaseous analyte enters a sampler and diffuses at a known rate through a barrier into a chamber where it is trapped on a sorbent. The sorbent is later analyzed to determine the amount of analyte present. The airborne concentration of the analyte is calculated by dividing the mass of sorbed analyte (ng) by the deployment time (days) and the sampling rate (m^3 day^{-1}). Passive air samplers (PAS) are increasingly being used for studies where spatially-resolved data are needed, or where active sampling is not possible due to cost, site restrictions, such as lack of electrical power or trained operators, or other constraints [20–23]. The main advantage is that a large number of samplers can be deployed to increase area coverage and improve spatial resolution. The main limitations are that the samplers require longer periods of time to collect the analyte, limiting temporal resolution, and, specifically for atmospheric Hg, that measurements of atmospheric mercury forms other than GEM (e.g., GOM and PBM) remain challenging, although some designs have had success [24].

The MerPAS® from Tekran is a commercially available mercury passive air sampler (PAS) that traps GEM on sulfur-impregnated activated carbon and uses a diffusive barrier to constrain the sampling rate [25]. The device includes a protective shield for deployment outdoors, where it can be left to collect GEM for months without revisiting the site until it is removed for analysis. At the laboratory the sorbent is analyzed, typically with a direct mercury analyzer (DMA), and the concentrations of GEM are calculated as discussed earlier; details of the entire method and sampling rate calculations are described in Section 2.

Recent research has shown that the MerPAS® sampler can not only measure GEM but also characterize and quantify atmospheric mercury sources, both with and without isotope tracing [26,27]. We have recently shown that the MerPAS® sampler can also discriminate landscape, seasonal, and elevation effects on GEM if given sufficient collection time, adequate analytical precision, and low blank levels [28]. In the present study, we used MerPAS® units to quantify GEM at six sites along the Mississippi and Alabama Gulf Coast during four consecutive seasons, from spring 2019 to winter 2020. Herein, we report our results with emphasize on spatial and temporal trends in GEM, and a comparison between passive and active sampling data co-collected at Grand Bay NERR, a National Atmospheric Deposition Program Atmospheric Mercury Network (AMNet) site.

2. Material and Methods

2.1. Study Sites and Meteorological Measurements

GEM was determined at five locations along the Mississippi Gulf Coast, including Bay St. Louis, Gulf Port, Gulf Coast Research Laboratory (GCRL) main campus near Ocean Springs, GCRL at Cedar Point, Grand Bay NERR near Moss Point, as well as at the Dauphin Island Sea Lab located to the southeast on a barrier island in Alabama (Figure 1). Figure 1 also shows anthropogenic Hg point sources based on US Environmental Protection Agency 2018 toxic release inventory (TRI) data, the most recent TRI data available [29]. Table A1 in Appendix A provides site coordinates, sampling periods, and mean temperature and wind speed during deployment. Meteorological data stems from the nearest weather stations, ranging from on-site at Grand Bay to 4.9 km away at Bay St. Louis. The samplers were deployed for ~4 weeks during 4 consecutive seasons, starting in spring 2019. The Grand Bay site has been described in detail elsewhere [17]. Briefly, the National Oceanic and Atmospheric Administration's (NOAA's) Air Resources Laboratory established Hg monitoring at the wetland site in 2006, and has been operating Tekran systems there as part of the National Atmospheric Deposition Program's AMNet. Long-term observations of atmospheric speciated Hg at the site have been published elsewhere [16]. The Cedar City site was also within a coastal wetland, whereas the other sites were at the immediate coastline with MerPAS® units deployed above open water.

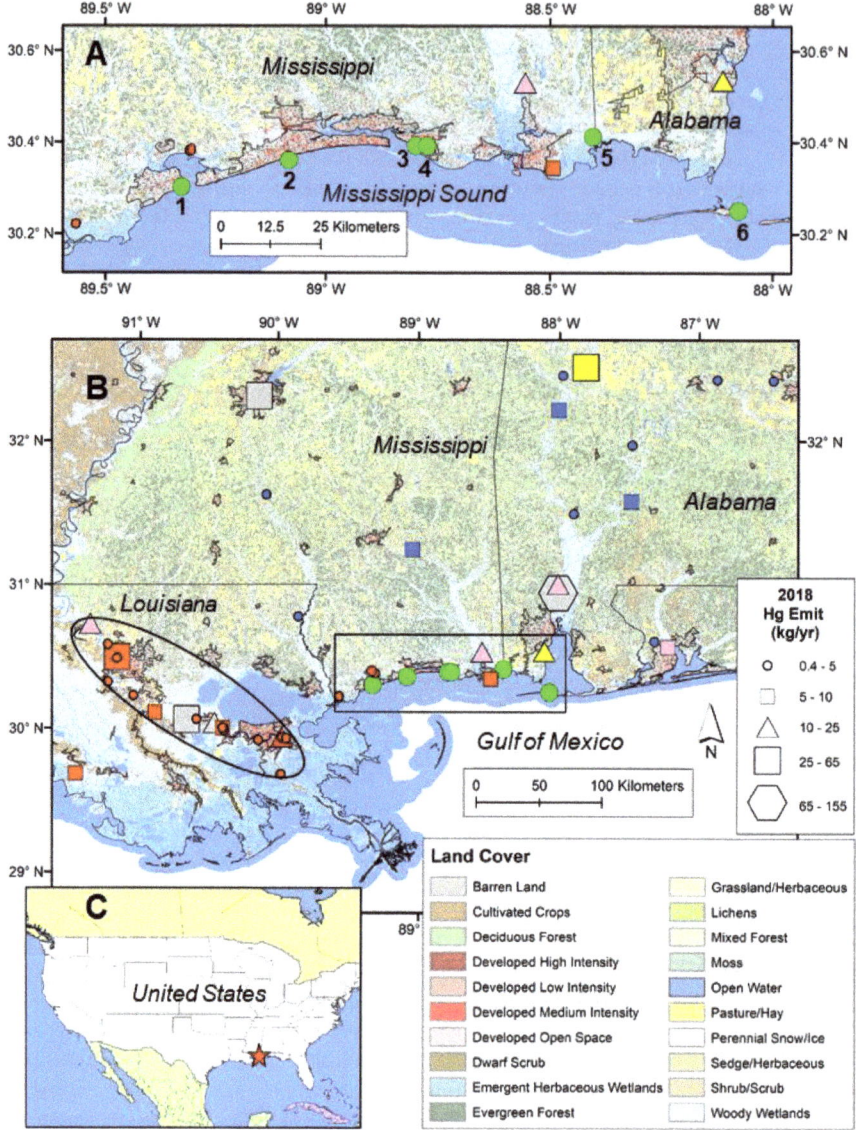

Figure 1. Maps of the study site. (**A**) = close up showing the six sampling locations as green circles (1 = Bay St. Louis; 2 = Gulfport; 3 = Gulf Coast Research Laboratory (GCRL) Main; 4 = GCRL Cedar Point; 5 = Grand Bay National Estuarine Research Reserve (NERR); 6 = Dauphin Island). (**B**) = a regional view with the close-up domain indicated in the box. (**C**) = a general location map showing the study site with star. The Hg air emission point sources are based on the most recent toxic release inventory data (2018) [29], where the size and shape of the emissions symbols indicate the amount of emissions (kg/year) and the color of the symbol indicates the source category: refineries and chemicals (red); electric power generation (pink); metals (gray); paper (blue); cement (yellow). Land cover categories are based on the 2011 National Land Cover Database [30]. The New Orleans and Baton Rouge area in Louisiana (shown with an oval) has gaseous elemental mercury (GEM) emission sources from multiple industries.

2.2. The MerPAS® and Its Preparation and Deployment in This Study

We used four to six MerPAS® units (Tekran Corp., Toronto, ON, Canada) to concurrently collect GEM at each site during each deployment (Figure 2). The development and performance characteristics of the passive sampler have been described in detail elsewhere [25,31,32]. Briefly, sulfur-impregnated activated carbon serves as a sorbent, and is housed in a stainless-steel mesh cylinder at the center of the device (Figure 1). The mesh is inserted into a microporous diffusive barrier (white Radiello®, Sigma Aldrich, St. Louis, MO, USA) which constrains the sampling rate. GEM diffuses through the barrier and is retained on the sorbent, but GOM and PBM do not appreciably pass the barrier [33]. The diffuse barrier itself is screwed into the center of a protective shield that permits outdoor deployment. The shield has an opening at the bottom that allows for air circulation but keeps precipitation out.

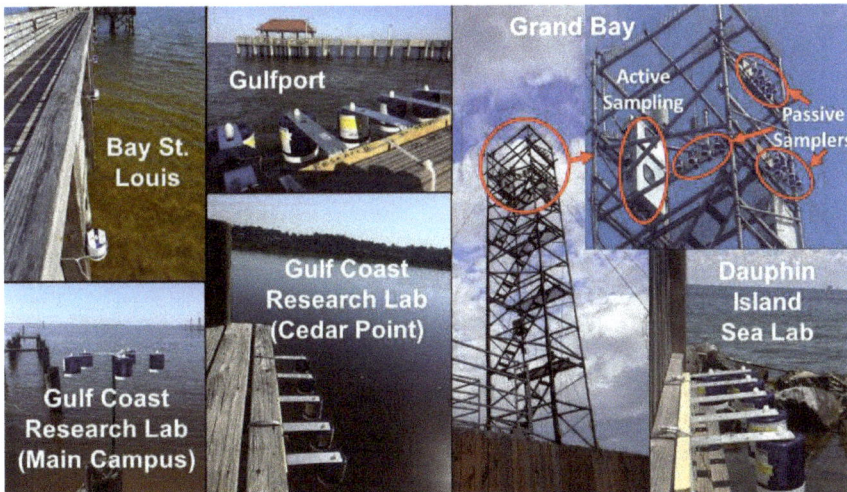

Figure 2. Views of the six sampling sites along the northern Gulf of Mexico (nGoM) showing deployment of passive air samplers for GEM collection.

In this study, we loaded about 0.6 g of freshly crushed and sieved (250–1000 μm) sulfur-impregnated activated carbon sorbent (HGR-AC, Calgon Carbon Corp., Pittsburg, PA, USA) into the samplers <3 days prior to deployment. Before loading the samplers, the activated carbon was analyzed for Hg to ascertain the blank level; the sorbent was only used when it would contribute <0.15 ng of Hg per sampler, amounting to <4% of the Hg accumulated during sampling. Samplers were deployed at 1.5 to 3.0 m above the water to prevent water from splashing into the device, except at Grand Bay where they were deployed at the top of a 10 m tower. We did not observe any salt inside the samplers and do not suspect water splashed into them. After each use, we cleaned diffusive barriers with a stream of nitrogen, and would only re-use them if they remained clean and undamaged; others have shown no significant difference in sampling rate between new and used barriers if the barriers are kept clean and in good condition [32].

2.3. Determination of Hg Collected on the Pas Sorbent and Calculation of Atmospheric Hg Concentratins

Upon retrieval the PASs were capped, sealed with polytetrafluoroethylene tape, placed in Ziplock bags, transported to the laboratory, and stored in a clean room until analysis within 2 days of collection. Details of the analysis were described in a previous study [28]. Briefly, total Hg collected on the sorbent was determined by a Direct Mercury Analyzer (DMA-80; Milestone Inc., Shelton, CT, USA), a technique which is based on thermal decomposition, gold amalgamation, and atomic absorption spectrometry. We followed U.S. Environmental Protection Agency (EPA) Method 7473, with some modifications for trapping the sulfur released from the sulfur-impregnated activate carbon with Na_2CO_3 [34]. Prior to analysis, quartz sample holders (boats) were pre-cleaned by soaking in 5% nitric acid overnight, rinsed with deionized water, and heated to 550 °C for several hours to remove any traces of Hg. Then the sorbent within the stainless-steel mesh cylinder was weighed into the boats and covered with ~0.2 g of Na_2CO_3. This process was repeated using a second boat because the capacity of a single boat was not enough for the amount of sorbent in a PAS. The boats were then loaded into the autosampler and analyzed by the DMA, with the Hg for the two boats being combined. The DMA instrument was calibrated using Hg solutions that were prepared from a 10 µg ml^{-1} Hg stock solution (Spex Certiprep, Metuchen, NJ, USA). Coal fly ash standard reference material (SRM; NIST 1633C) was analyzed before beginning the sample analysis and every 20 boats thereafter. Recovery of SRM over the analyses was 94.6 ± 4.2% (mean ± SD, $n = 16$). The limit of detection was 0.014 ng of Hg.

Concentrations of GEM were calculated by dividing the mass of adsorbed Hg (ng) by the deployment time (days) and the sampling rate (m^3 day^{-1}). Hg uptake (after blank subtraction) ranged from 3.14 to 4.58 ng during ~4 weeks deployment period. We used a sampling rate of 0.111 m^3 day^{-1} recommended by Tekran. The sampling rate was adjusted for local temperature and wind speed, factors which can influence the molecular diffusivity of GEM and the overall sampling rate, respectively [32]. Meteorological data are given in Table A1 in Appendix A. The adjusted sampling rate was calculated using Equation (1) [32] and ranged from 0.106 to 0.133 (m^3 day^{-1}), depending on season and location.

$$SR_{adj.} = SR_{cal} + (T - 9.89\,°C) \cdot 0.0009\ m^3\ (day\ °C)^{-1} + (W - 3.41\ m\ s^{-1}) \cdot 0.003\ m^3\ (day\ °C)^{-1} \quad (1)$$

2.4. Measurement of GEM at NOAA's Grand Bay Site Using Active Sampling

Atmospheric speciated mercury (GEM, GOM, and PBM) was monitored at the Grand Bay using a Tekran speciation system, which has been described elsewhere [12,17]. Briefly, ambient air is sampled by the mercury detection system at approximately 10 L/min. Large particles (d > 2.5 µm) are removed at the inlet by an elutriator/impactor assembly, GOM is collected on a KCl-coated quartz annular denuder, and PBM (d < 2.5 µm) is collected on a quartz regenerable particle filter (RPF). GEM passes though the glassware unimpeded and is sequentially collected on one of two gold traps at 5-min intervals. As one trap collects GEM, the other is heated to thermally desorb GEM into a flow of argon, and the liberated GEM is detected via cold vapor atomic fluorescence spectrometry. After one hour, sample collection ceases and the collected GOM on the denuder and PBM on the quartz filter are then sequentially thermally desorbed in a flow of mercury-free zero air and quantitatively converted to GEM, which is then analyzed by the mercury detector. Thus, the speciation system operates on a 50% duty cycle, and reports GEM in real time at 5-min intervals during the sampling hour, and one-hour integrated concentrations of GOM and PBM during the subsequent desorption cycle. AMNet standard operating protocols ([35], http://nadp.slh.wisc.edu/AMNet/docs.aspx) were followed for mercury measurement and data reduction. Herein we focus on the GEM data for comparison with our PAS data.

2.5. Statistical Analysis

Differences in passively sampled GEM concentrations among locations and seasons and the interaction between location and season were examined using univariate repeated measures analysis of variance (rmANOVA). Location was treated as a between-subjects effect, whereas season and its interaction with location were treated as within-subjects effects. Given a significant main effect of location, Tukey's tests of honest significant differences (HSD tests) were used to examine pairwise differences in GEM means among locations. Pairwise differences among seasons were tested using t-tests and a Sidak p-value adjustment for multiple comparisons. The components of a significant season · location interaction were tested using planned orthogonal contrasts. Contrasts were chosen to test the hypothesis that GEM concentrations were greater outside the growing season (winter) than during other times of the year, and lower during the hottest growing season (summer) than in the spring and fall. Contrasts associated with the components of the season · location interaction included (1) the difference in GEM between winter and the remaining seasons depended on location ((winter v. rest) · location), (2) the difference in GEM between summer and the average of spring and fall depended on location ((summer v. spr/fall) · location), and (3) the difference in GEM between spring and fall depended on location ((spr v. sum) · location). Differences were deemed significant at a $p < 0.05$ level.

Four one-sample t-tests (one for each season) were used to examine GEM concentration differences between active and passive sampling methods at Grand Bay NERR. Since the same active sampler was used to take hundreds of measurements in a given season, the measurements could not be considered independent observations. Hence, we averaged all actively sampled measurements in each season, assuming no replication and thus no within-season variation for the active sampler. We then compared the sample mean and standard error (SE, defined as the sample standard deviation (SD) divided by the $\sqrt{n}$) of passively-sampled GEM concentrations for the deployment periods in each season to the average GEM concentration for the active sampler during the corresponding period. Although the one-sample t-tests assumed no statistical error, active samplers have an estimated 10% measurement error [16,17]. We therefore assumed that the seasonal average measurement of GEM by each active sampler represented the midpoint of this 10% measurement uncertainty interval. We corrected the p-values produced by each of the four one-sample t-tests using Sidak's multiplicative correction for multiple t-tests. Differences were deemed significant at a $p < 0.05$ level. Data were analyzed using SYSTAT (version 13.0, San Jose, CA, USA).

3. Results and Discussion

Overall precision between the samplers deployed side-by-side averaged ~7% relative standard deviation, which is in the expected range for this method [31]. Adjustment of the sampling rate using local meteorological data generally decreased the GEM levels from 0–8%, except during the cold fall and winter periods where GEM increased at a few sites by up to 6%, and during July 2019 when a windy summer tropical storm helped decrease GEM levels by as much as 14% at Gulfport. Adjustments at Grand Bay were generally greater because it tended to be windier atop the 10 m tower.

3.1. Seasonal Trends of GEM Concentration along the nGoM Using PASs

GEM concentrations (ng m^{-3}) varied significantly among seasons (rmANOVA $F_{season\ 3,75} = 107.58$; unadjusted and Greenhouse–Geisser $p << 0.01$). Mean seasonal GEM concentrations (ng m^{-3} ± 1 SE) at each of the six locations are shown in Figure 3, with specific values given in Table A2 in Appendix A. Mean concentrations ranged from 1.00 ± 0.03 ng m^{-3} in the summer at GCRL Cedar Point to 1.77 ± 0.03 ng m^{-3} in the winter at Bay St. Louis. Differences among seasons averaged across locations revealed that GEM concentrations were significantly higher in the winter than in all other seasons across all sites (1.53 ± 0.03 (winter), 1.18 ± 0.03 (fall), 1.25 ± 0.03 (spring), and 1.14 ± 0.02 (summer); Sidak-adjusted $p < 0.05$; Figure 3). GEM concentrations were lower in the summer than in spring (Sidak $p < 0.01$; Figure 3), but not significantly lower than in fall (Sidak-adjusted $p = 0.12$). There was no significant difference in GEM concentration between spring and fall (Sidak-adjusted $p = 0.29$; Figure 3).

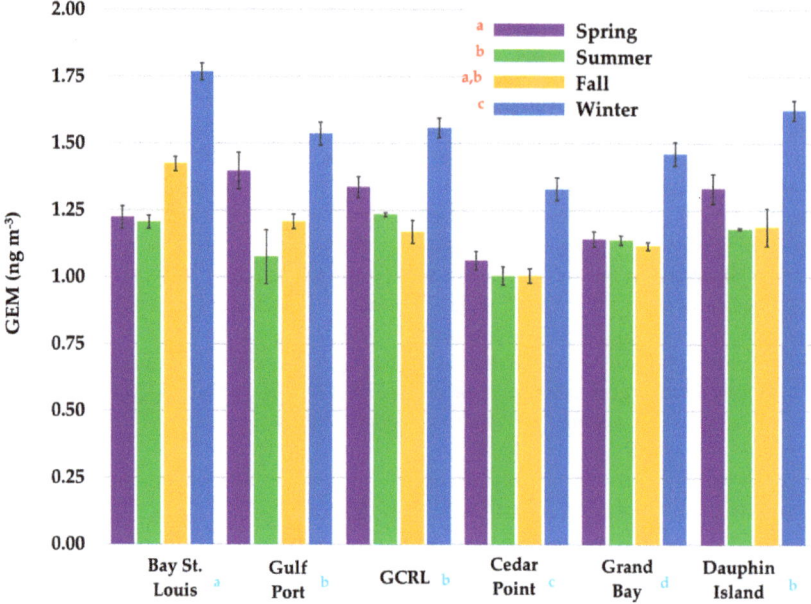

Figure 3. GEM concentrations determined using passive air samplers (PASs) deployed at six sites along the nGoM from May 2019 to February 2020. Sites are depicted west to east (from left to right) and error bars represent 1 standard error. Results for two sets of statistical analyses are shown: (1) pairwise means comparisons for the main effect of season (red letters), and (2) pairwise means comparisons for the main effect of location (blue letters). Seasons or locations that do not share letters are statistically different ($p < 0.05$) as determined by Tukey's honest significant difference tests. The season · location interaction is described in the text.

GEM levels tend to be higher in the winter due in part to the uptake of Hg by plants during the growing season which generally extends from spring through early fall [5], but also due to shifts in prevailing winds which are generally from the south (arriving from the GoM) in the summer and from the north (over terrestrial areas with point sources) in the winter (Figure 2) [16,17]. Other factors that can contribute to seasonal differences in atmospheric Hg species include greater sunlight intensity in the summer, which can increase conversion of GEM to GOM by photochemical oxidation, and precipitation in the summer from convective thunderstorms that can strip GOM from the air, resulting in high levels of wet Hg deposition [16,17,36]. Seasonal trends of airborne Hg species in southeastern U.S. have now been studied using both active and passive sampling, and our data are consistent with previously reported trends [16,37,38].

The pattern of seasonal differences in GEM concentrations varied among locations, as indicated by a significant season x location interaction (rmANOVA $F_{season \cdot location\ (15,75)} = 3.14$; unadjusted and Greenhouse–Geisser $p < 0.01$). The difference in GEM between fall and spring varied among locations (Contrast $F_{fall\ v.\ spring \cdot location\ (5,25)} = 3.14$; $p = 0.03$). Whereas GEM in the fall was greater than GEM in the spring at Bay St. Louis, the same was not true at other locations (Figure 3). The difference in GEM between summer and the average of spring and fall varied significantly among locations (Contrast $F_{summer\ v.\ spr/fall \cdot location\ (5,25)} = 4.12$; $p < 0.01$). Whereas GEM was lower in the summer than the average for spring or fall at Bay St. Louis and Gulfport, this difference was lower at Dauphin Island, Cedar Point, and GCRL, and absent at Grand Bay (Figure 3). Spatial differences are examined further below.

3.2. Spatial Trends of GEM Concentration along the nGoM Using PASs

Previous studies have shown that coastal sites can be influenced by both polluted air from urban environments and cleaner Gulf of Mexico marine air [17,39]. In our study, GEM concentrations varied significantly among locations (rmANOVA $F_{location\,(5,25)}$ = 38.60; $p \ll 0.01$). Averaged across seasons, Tukey's HSD tests revealed that GEM at Bay St. Louis was higher than at all other sites ($p < 0.05$) (Figure 3). As the western-most site, Bay St. Louis is closest to New Orleans (<100 km), by far the largest population center in the area with a number of Hg sources from various industries. For the New Orleans and Baton Rouge area, Hg emissions in 2018 amounted to ~206 kg, more than double the amount emitted from all the sites in Mississippi shown in Figure 1. In addition, there is a close-in point source ~6 km to the north of the Bay St. Louis site (Figure 1). Generally there are higher GEM concentrations from the north and northeast during the winter and from the southwest during the summer for active sampling at the Grand Bay site (Figure 4). Detailed air mass back-trajectory and source-receptor modelling at each site is beyond the scope of this work. Additional study is needed to determine the persistence and cause of the higher GEM concentrations found at Bay St. Louis.

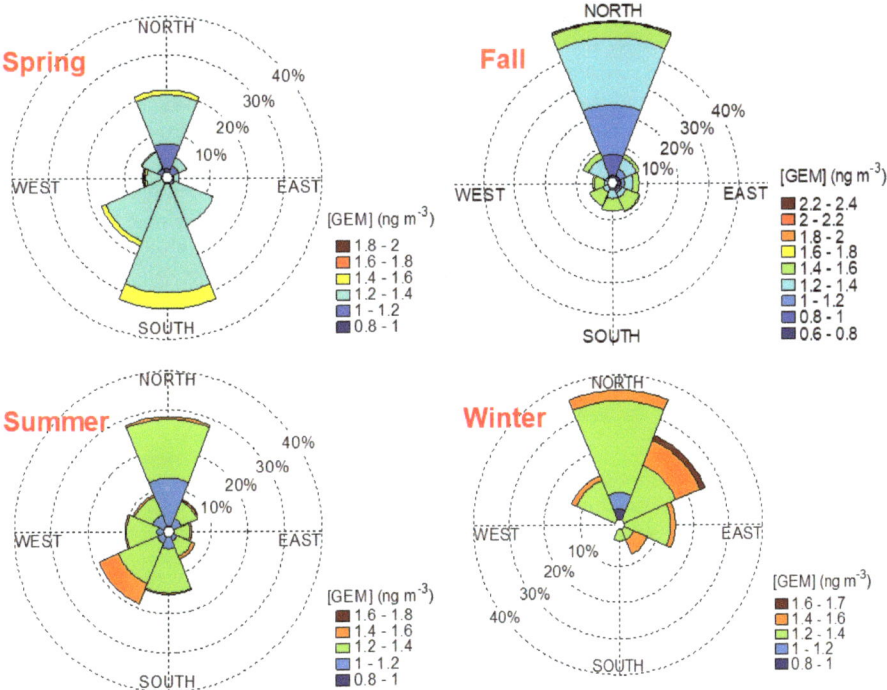

Figure 4. Wind roses showing the relationship between GEM levels and wind direction for each sampling period at Grand Bay.

We also observed the lowest GEM concentrations at the Cedar Point coastal marsh site (Figure 3). The site is located away from the coastal beach area in a sheltered Bayou and was in proximity to the most surrounding vegetation, a known sink for airborne Hg. The Grand Bay site is also within a wetland, but we sampled there from the top of the 10 m tower, likely capturing air masses relatively unimpeded by vegetation, which may have moderated the wetland effect. Tukey's HSD tests revealed no statistical differences among the other open water coastal sites (Gulfport, GCRL Main Campus, and Dauphin Island; $p > 0.83$). However, there are certainly additional complexities in this sub-tropical coastal environment that passive air samplers are unable to resolve given their long deployment times.

For example, although GOM data is not included herein, GOM concentrations at coastal sites can be influenced not only by regional point sources [16], but by conversion of GEM to GOM through photochemical oxidation associated with halogen species, such as BrO and BrCl, derived from marine aerosols [40–42].

3.3. Comparison of GEM Determined by Cctive and Passive Sampling at Grand Bay

Compared to passive sampling, active sampling provides high temporal resolution with many more data points. At Grand Bay, we observed diurnal variations, seasonal trends, unknown plume events, and other complexity (Figure 3). Detailed analysis of Hg species fluctuations is beyond the scope of this study, but others have reported on this in the region [16,39,43]. Here, we focus on preliminary data comparing GEM concentrations between passive and active sampling techniques for data co-collected at the AMNet Grand Bay NERR site. It is worth mentioning that GEM levels have been declining at the Grand Bay site at a rate of -0.009 ng m^{-3}/yr from 2007–2018, which may be partly explained by a concurrent decrease in anthropogenic Hg emissions in the region, especially for the electric power generating industry [16,29].

Summary statistics for GEM concentrations determined at Grand Bay are given in Table 3 in Appendix A. We observed similar seasonal trends in GEM concentration with highest concentrations in winter by both active and passive sampling methods. However, active sampling gave slightly higher mean GEM concentrations in the spring, summer, and fall, but not in the winter (one-sample $t_{(passive-active)}$ = −5.16, −7.13, −10.66, and 1.43 for spring, summer, fall, and winter, respectively, with df = 5, 5, 11, and 5; Figure 5). The trend is also depicted in Figure 3. It is unclear why passive sampling gave slightly lower average concentrations compared to active sampling for the spring, summer, and fall, and why winter was the exception. The re-use of the passive samplers may have caused a small bias or the activated carbon stock may have changed in some way over time, although it was still analyzed prior to analyses for blank subtraction. Nevertheless, the <15% difference between the averages of the two methods, operated by two different groups, may be considered acceptable, especially when evaluating larger spatial and temporal trends.

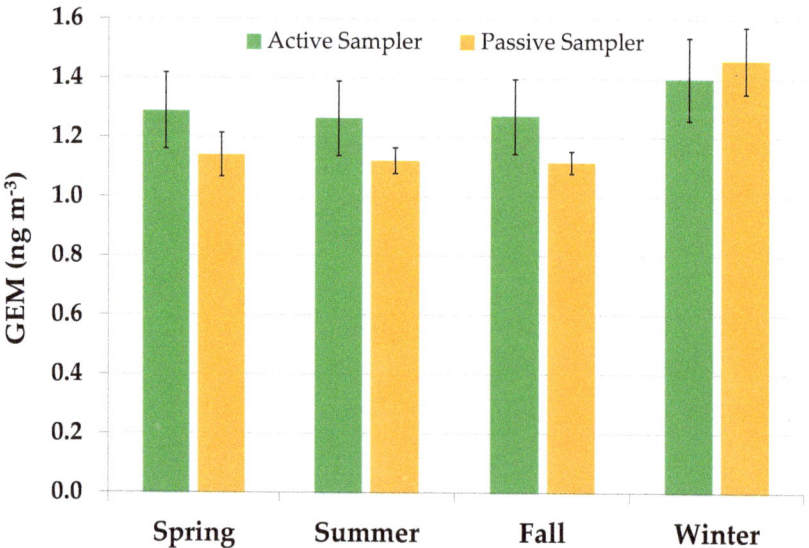

Figure 5. GEM concentrations determined using passive and active sampling at Grand Bay NERR. Error bars for passive sampler data represent 95% confidence intervals. Bars for the active sampler data represent 10% measurement error.

4. Conclusions

We deployed MerPAS® passive air samplers to determine GEM at multiple sites along the nGoM over the course of a year. We observed higher GEM levels in the winter compared to other seasons across the sites. Spatially, mean GEM levels were highest at Bay St. Louis, the western-most site nearest New Orleans, and lowest at Cedar Point, a coastal marsh site with extensive vegetation. MerPAS® units were also deployed at Grand Bay near a Tekran air Hg speciation system that is based on active sampling. The passive air samplers gave slightly lower concentrations to the active sampling method, except in the winter. Despite the difference, the MerPAS® passive air samplers were capable of discriminating both seasonal and spatial differences, providing further insight into the sources and factors that influence GEM along the nGoM.

Author Contributions: J.V.C. conceptualized, supervised, and administered the project; B.J. led the P.A.S. sampling and measurement campaign; J.S.B. and B.J. provided statistically analyses; B.J. and J.V.C. prepared the original draft; X.R., P.K. and W.T.L. provided the Grand Bay active sampling data; M.D.C. prepared Figure 1; X.R. prepared the wind rose plots. All authors helped review and edit the paper. All authors have read and agreed to the published version of the manuscript.

Funding: This research was funded by seed grants from the Mississippi Space Grant Consortium and RII Track-2 FEC: Emergent Polymer Sensing Technologies for Gulf Coast Water Quality Monitoring, which was part of NSF Award #1632825.

Acknowledgments: We thank those who gave us permission to deploy our PASs at our sampling sites. We are particularly grateful to Mike Archer at the Grand Bay NERR for access to the AMNet tower and help with sampling. We are also thankful to Eric Prestbo, Lucas Hawkins, Diana Babi, and others at Tekran Corp. for their continued advice and support.

Conflicts of Interest: The authors declare no conflict of interest.

Appendix A

Table A1. Coordinates for sampling sites and nearest weather stations, along with sampling periods and mean temperature and wind speed during deployment.

	Bay St. Louis		Gulf Port		GCRL (Main Campus)		GCRL (Cedar Point)		Grand Bay		Dauphin Island	
Deployment Sites:	30.302°N, 89.327°W		30.361°N, 89.083°W		30.392°N, 88.799°W		30.392°N, 88.775°W		30.412°N, 88.404°W		30.251°N, 88.077°W	
Weather Stations:	30.287°N, 89.376°W		30.364°N, 89.086°W		30.401°N, 88.808°W		30.401°N, 88.773°W		30.412°N, 88.404°W		30.254°N, 88.103°W	
Sampling Period	Temp. (°C)	Wind (m/s)	Temp. (°C)	Wind (m/s)	Temp. (°C)	Wind (m/s)	Temp. (°C)	Wind (m/s)	Temp. (°C)	Wind (m/s)	Temp. (°C)	Wind (m/s)
May–June (16/5/2019–13/6/2019)	27.0	1.6	27.0	5.7	27.2	0.3	26.4	0.9	26.7	3.0	27.2	1.0
June–July (13/6/2019–11/7/2019)	28.1	1.4	28.5	4.7	28.7	0.3	27.9	0.9	27.9	2.9	28.7	1.2
July–August (11/7/2019–8/8/2019)	26.8	1.3	26.4	4.9	27.4	0.3	26.8	0.8	26.7	2.2	28.3	1.1
August–September (8/8/2019–5/9/2019)	27.4	0.8	28.4	3.5	27.5	0.2	27.1	0.4	27.4	1.8	28.5	0.9
November–December (1/11/2019–3/12/2019)	12.8	1.0	13.7	3.6	13.0	0.5	13.0	0.5	13.4	2.3	14.9	2.1
January–February (27/1/2020–18/2/2020)	13.7	1.3	14.3	4.6	14.3	0.7	14.3	1.1	14.8	4.9	14.6	2.1

Table A2. Amount of Hg (ng) collected on each PAS and GEM concentrations based on those amounts ($n = 6$, unless otherwise noted).

Sampling Period	Amount of Hg Collected and GEM Level	Bay St. Louis Mean	SE	Gulf Port Mean	SE	GCRL Main Campus Mean	SE	GCRL Cedar Point Mean	SE	Grand Bay Mean	SE	Dauphin Island Mean	SE	All Locations Mean	SE
16/5/2019–13/6/2019	Hg (ng)	4.12 [a]	0.14	4.60	0.22	4.54	0.13	3.64	0.12	4.00	0.11	4.48	0.19	4.23	0.15
	Conc. (ng m^{-3})	1.22	0.04	1.40	0.07	1.34	0.04	1.06	0.03	1.14	0.03	1.33	0.06	1.25	0.03
13/6/2019–11/7/2019	Hg (ng)	3.77 [b]	0.09	3.90	0.08	3.95	0.17	3.45 [a]	0.13	3.77	0.12	3.85	0.06	3.78	0.07
	Conc. (ng m^{-3})	1.15	0.03	1.10	0.02	1.24	0.05	1.06	0.04	1.11	0.04	1.18	0.02	1.14	0.02
11/7/2019–8/8/2019	Hg (ng)	Lost in tropical storm		3.36	0.21	4.07	0.11	3.14 [a]	0.06	4.00	0.16	3.98 [a]	0.12	3.71	0.70
	Conc. (ng m^{-3})			0.89	0.05	1.24	0.03	0.95	0.02	1.17	0.05	1.18	0.04	1.09	0.05
8/8/2019–5/9/2019	Hg (ng)	4.19 [a]	0.07	4.29	0.06	3.95	0.10	3.25	0.21	3.98	0.07	4.11	0.02	3.96	0.15
	Conc. (ng m^{-3})	1.26	0.02	1.24	0.02	1.22	0.03	1.00	0.06	1.13	0.02	1.19	0.01	1.17	0.02
1/11/2019–3/12/2019	Hg (ng)	4.36 [a]	0.08	4.43 [a]	0.10	3.92 [a]	0.14	3.37 [a]	0.09	3.96	0.05	4.26	0.25	4.05	0.16
	Conc. (ng m^{-3})	1.42	0.03	1.21	0.03	1.17	0.04	1.00	0.03	1.12	0.01	1.19	0.07	1.18	0.03
27/1/2020–18/2/2020	Hg (ng)	4.58 [b]	0.08	4.35	0.12	4.17	0.10	3.60	0.12	4.42	0.13	4.50	0.10	4.27	0.15
	Conc. (ng m^{-3})	1.77	0.03	1.54	0.04	1.56	0.04	1.33	0.04	1.46	0.04	1.62	0.04	1.53	0.03
All Seasons	Hg (ng)	4.21	0.07	4.15	0.09	4.10	0.06	3.42	0.06	4.02	0.05	4.20	0.07		
	Conc. (ng m^{-3})	1.36	0.05	1.23	0.04	1.29	0.03	1.07	0.09	1.19	0.02	1.28	0.03		

[a] $n = 5$; [b] $n = 4$. SE = Standard Error

Table 3. Summary statistics for GEM concentrations at Grand Bay NERR by active and passive sampling along with meteorological data used to obtain the adjusted sampling rate for each PAS.

Season	Mean Temperature (°C)	Mean Wind Speed (m/s)	Statistical Parameter	Active Sampler (ng m^{-3})	Passive Sampler (ng m^{-3})
Spring 2019	26.7	3.0	n Range Mean Median SD	324 0.90–1.81 1.29 1.30 0.10	6 1.07–1.27 1.14 1.14 0.07
Summer 2019	27.3	2.5	n Range Mean Median SD	550 0.98–1.64 1.26 1.26 0.10	18 1.03–1.38 1.14 1.12 0.09
Fall 2019	13.4	2.3	n Range Mean Median SD	371 0.71–1.68 1.27 1.30 0.20	6 1.06–1.15 1.12 1.12 0.04
Winter 2020	14.8	4.9	n Range Mean Median SD	256 0.89–1.66 1.40 1.39 0.14	6 1.35–1.64 1.46 1.43 0.11

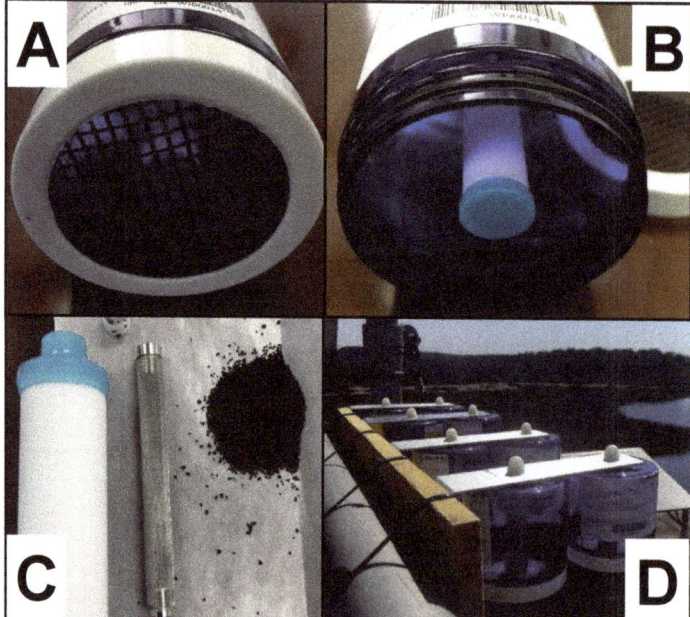

Figure 1. Photos showing the MerPAS® configuration with cover on (**A**), with the cover off (**B**), and with diffusive body, stainless steel screen, and activated carbon sorbent removed (**C**), and deployment on the tower at Grand Bay (**D**).

Atmosphere 2020, 11, 1034

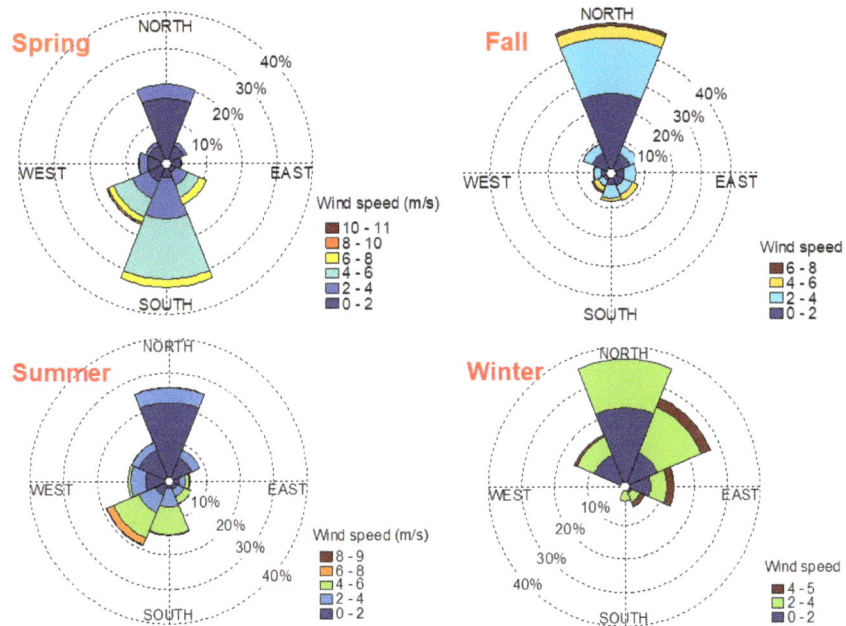

Figure 2. Wind roses showing the relationship between wind speed and wind direction for each sampling period at Grand Bay.

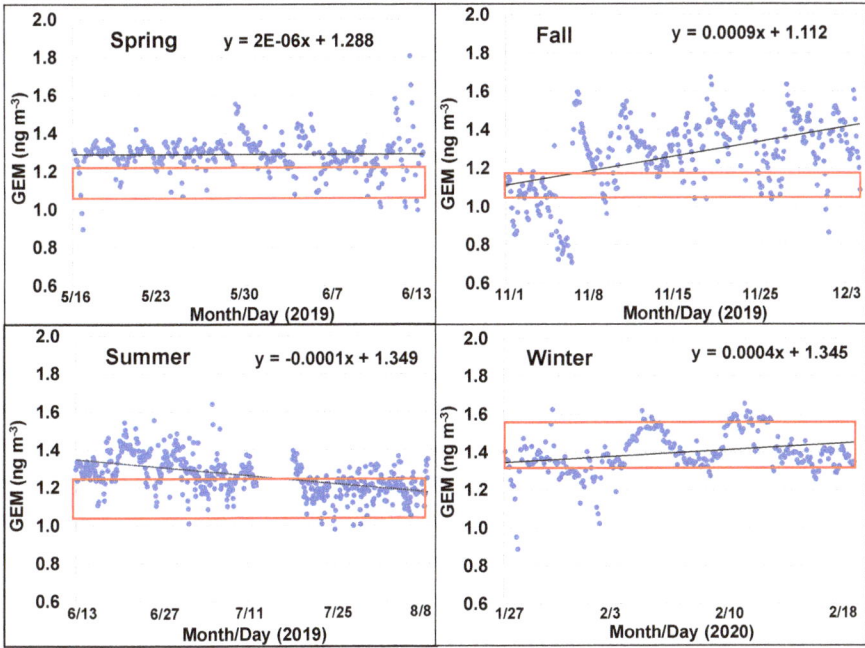

Figure 3. Hourly GEM concentrations determined at the Grand Bay NERR site using active sampling. The red box encompasses the passive sampler data (average ± 1SD) obtained for the same period. The equation is for the linear regression of the data with the trend line in black.

References

1. Fitzgerald, W.F.; Engstrom, D.R.; Mason, R.P.; Nater, E.A. The Case for Atmospheric Mercury Contamination in Remote Areas. *Environ. Sci. Technol.* **1998**, *32*, 1–7. [CrossRef]
2. Gustin, M.; Jaffe, D. Reducing the Uncertainty in Measurement and Understanding of Mercury in the Atmosphere. *Environ. Sci. Technol.* **2010**, *44*, 2222–2227. [CrossRef] [PubMed]
3. Lyman, S.N.; Cheng, I.; Gratz, L.E.; Weiss-Penzias, P.; Zhang, L. An updated review of atmospheric mercury. *Sci. Total Environ.* **2020**, *707*, 135575. [CrossRef] [PubMed]
4. Gustin, M.S.; Amos, H.M.; Huang, J.; Miller, M.B.; Heidecorn, K. Measuring and modeling mercury in the atmosphere: A critical review. *Atmos. Chem. Phys.* **2015**, *15*, 5697–5713. [CrossRef]
5. Schroeder, W.H.; Munthe, J. Atmospheric mercury—An overview. *Atmos. Environ.* **1998**, *32*, 809–822. [CrossRef]
6. Driscoll, C.T.; Mason, R.P.; Chan, H.M.; Jacob, D.J.; Pirrone, N. Mercury as a Global Pollutant: Sources, Pathways, and Effects. *Environ. Sci. Technol.* **2013**, *47*, 4967–4983. [CrossRef] [PubMed]
7. Lin, C.-J.; Pehkonen, S.O. The chemistry of atmospheric mercury: A review. *Atmos. Environ.* **1999**, *33*, 2067–2079. [CrossRef]
8. Sprovieri, F.; Pirrone, N.; Bencardino, M.; D'Amore, F.; Carbone, F.; Cinnirella, S.; Mannarino, V.; Landis, M.; Ebinghaus, R.; Weigelt, A.; et al. Atmospheric mercury concentrations observed at ground-based monitoring sites globally distributed in the framework of the GMOS network. *Atmos. Chem. Phys.* **2016**, *16*, 11915–11935. [CrossRef]
9. Zhang, Y.; Jacob, D.J.; Horowitz, H.M.; Chen, L.; Amos, H.M.; Krabbenhoft, D.P.; Slemr, F.; St. Louis, V.L.; Sunderland, E.M. Observed decrease in atmospheric mercury explained by global decline in anthropogenic emissions. *Proc. Natl. Acad. Sci. USA* **2016**, *113*, 526–531. [CrossRef]
10. Choi, A.L.; Grandjean, P. Methylmercury exposure and health effects in humans. *Environ. Chem.* **2008**, *5*, 112–120. [CrossRef]
11. Hall, B.D.; Aiken, G.R.; Krabbenhoft, D.P.; Marvin-DiPasquale, M.; Swarzenski, C.M. Wetlands as principal zones of methylmercury production in southern Louisiana and the Gulf of Mexico region. *Environ. Pollut.* **2008**, *154*, 124–134. [CrossRef] [PubMed]
12. Ren, X.; Luke, W.T.; Kelley, P.; Cohen, M.D.; Artz, R.; Olson, M.L.; Schmeltz, D.; Puchalski, M.; Goldberg, D.L.; Ring, A.; et al. Atmospheric mercury measurements at a suburban site in the Mid-Atlantic United States: Inter-annual, seasonal and diurnal variations and source-receptor relationships. *Atmos. Environ.* **2016**, *146*, 141–152. [CrossRef]
13. Engle, M.A.; Tate, M.T.; Krabbenhoft, D.P.; Kolker, A.; Olson, M.L.; Edgerton, E.S.; DeWild, J.F.; McPherson, A.K. Characterization and cycling of atmospheric mercury along the central US Gulf Coast. *Appl. Geochem.* **2008**, *23*, 419–437. [CrossRef]
14. Lincoln, R.A.; Shine, J.P.; Chesney, E.J.; Vorhees, D.J.; Grandjean, P.; Senn, D.B. Fish Consumption and Mercury Exposure among Louisiana Recreational Anglers. *Environ. Health Perspect.* **2011**, *119*, 245–251. [CrossRef]
15. Merritt, K.A.; Amirbahman, A. Mercury methylation dynamics in estuarine and coastal marine environments—A critical review. *Earth Sci. Rev.* **2009**, *96*, 54–66. [CrossRef]
16. Ren, X.; Luke, W.T.; Kelley, P.; Cohen, M.D.; Olson, M.L.; Walker, J.; Cole, R.; Archer, M.; Artz, R.; Stein, A.A. Long-Term Observations of Atmospheric Speciated Mercury at a Coastal Site in the Northern Gulf of Mexico during 2007–2018. *Atmosphere* **2020**, *11*, 268. [CrossRef]
17. Ren, X.; Luke, W.; Kelley, P.; Cohen, M.; Ngan, F.; Artz, R.; Walker, J.; Brooks, S.; Moore, C.; Swartzendruber, P.; et al. Mercury Speciation at a Coastal Site in the Northern Gulf of Mexico: Results from the Grand Bay Intensive Studies in Summer 2010 and Spring 2011. *Atmosphere* **2014**, *5*, 230–251. [CrossRef]
18. Rolison, J.M.; Landing, W.M.; Luke, W.; Cohen, M.; Salters, V.J.M. Isotopic composition of species-specific atmospheric Hg in a coastal environment. *Chem. Geol.* **2013**, *336*, 37–49. [CrossRef]
19. Pandey, S.K.; Kim, K.-H.; Brown, R.J.C. Measurement techniques for mercury species in ambient air. *TrAC Trends Anal. Chem.* **2011**, *30*, 899–917. [CrossRef]
20. Gustin, M.S.; Lyman, S.N.; Kilner, P.; Prestbo, E. Development of a passive sampler for gaseous mercury. *Atmos. Environ.* **2011**, *45*, 5805–5812. [CrossRef]

21. Skov, H.; Sørensen, B.T.; Landis, M.S.; Johnson, M.S.; Sacco, P.; Goodsite, M.E.; Lohse, C.; Christiansen, K.S. Performance of a new diffusive sampler for Hg0 determination in the troposphere. *Environ. Chem.* **2007**, *4*, 75–80. [CrossRef]
22. Brumbaugh, W.G.; Petty, J.D.; May, T.W.; Huckins, J.N. A passive integrative sampler for mercury vapor in air and neutral mercury species in water. *Chemosphere Glob. Change Sci.* **2000**, *2*, 1–9. [CrossRef]
23. Peterson, C.; Alishahi, M.; Gustin, M.S. Testing the use of passive sampling systems for understanding air mercury concentrations and dry deposition across Florida, USA. *Sci. Total Environ.* **2012**, *424*, 297–307. [CrossRef] [PubMed]
24. Lyman, S.N.; Gustin, M.S.; Prestbo, E.M. A passive sampler for ambient gaseous oxidzied mercury concentrations. *Atmos. Environ.* **2010**, *44*, 246–252. [CrossRef]
25. McLagan, D.S.; Mitchell, C.P.J.; Huang, H.; Lei, Y.D.; Cole, A.S.; Steffen, A.; Hung, H.; Wania, F. A High-Precision Passive Air Sampler for Gaseous Mercury. *Environ. Sci. Technol. Lett.* **2016**, *3*, 24–29. [CrossRef]
26. Szponar, N.; McLagan, D.S.; Kaplan, R.J.; Mitchell, C.P.J.; Wania, F.; Steffen, A.; Stupple, G.W.; Monaci, F.; Bergquist, B.A. Isotopic Characterization of Atmospheric Gaseous Elemental Mercury by Passive Air Sampling. *Environ. Sci. Technol.* **2020**, *54*, 10533–10543. [CrossRef] [PubMed]
27. McLagan, D.S.; Monaci, F.; Huang, H.; Lei, Y.D.; Mitchell, C.P.J.J.; Wania, F. Characterization and Quantification of Atmospheric Mercury Sources Using Passive Air Samplers. *J. Geophys. Res. Atmos.* **2019**, *124*, 2351–2362. [CrossRef]
28. Jeon, B.; Cizdziel, J.V. Can the MerPAS Passive Air Sampler Discriminate Landscape, Seasonal, and Elevation Effects on Atmospheric Mercury? A Feasibility Study in Mississippi, USA. *Atmosphere* **2019**, *10*, 617. [CrossRef]
29. USEPA. Toxic Release Inventory. 2020. Available online: https://www.epa.gov/toxics-release-inventory-triprogram/tri-basic-data-files-calendar-years-1987-2018 (accessed on 13 January 2020).
30. ESRI. *USA National Land Cover Database 2011, Based on Data from the Multi-Resolution Land Characteristics Consortium*; ESRI ArcGIS Data Server: Redlands, CA, USA, 2019.
31. McLagan, D.S.; Mitchell, C.P.J.; Steffen, A.; Hung, H.; Shin, C.; Stupple, G.W.; Olson, M.L.; Luke, W.T.; Kelley, P.; Howard, D.; et al. Global evaluation and calibration of a passive air sampler for gaseous mercury. *Atmos. Chem. Phys.* **2018**, *18*, 5905–5919. [CrossRef]
32. McLagan, D.S.; Mitchell, C.P.J.; Huang, H.; Abdul Hussain, B.; Lei, Y.D.; Wania, F. The effects of meteorological parameters and diffusive barrier reuse on the sampling rate of a passive air sampler for gaseous mercury. *Atmos. Meas. Tech.* **2017**, *10*, 3651–3660. [CrossRef]
33. Stupple, G.; McLagan, D.; Steffen, A. In situ reactive gaseous mercury uptake on radiello diffusive barrier, cation exchange membrane and teflon filter membranes during atmospheric mercury depletion events. In Proceedings of the 14th International Conference on Mercury as a Global Pollutant (ICMGP), Krakow, Poland, 8–13 September 2019.
34. McLagan, D.S.; Huang, H.; Lei, Y.D.; Wania, F.; Mitchell, C.P. Application of sodium carbonate prevents sulphur poisoning of catalysts in automated total mercury analysis. *Spectrochim. Acta Part B Spectrosc.* **2017**, *133*, 60–62. [CrossRef]
35. Gay, D.A.; Schmeltz, D.; Prestbo, E.; Olson, M.; Sharac, T.; Tordon, R. The Atmospheric Mercury Network: Measurement and initial examination of an ongoing atmospheric mercury record across North America. *Atmos. Chem. Phys.* **2013**, *13*, 11339–11349. [CrossRef]
36. National Atmospheric Deposition Program/Mercury Deposition Network. Available online: http://nadp.slh.wisc.edu (accessed on 24 September 2020).
37. Nair, U.S.; Wu, Y.; Walters, J.; Jansen, J.; Edgerton, E.S. Diurnal and seasonal variation of mercury species at coastal-suburban, urban, and rural sites in the southeastern United States. *Atmos. Environ.* **2012**, *47*, 499–508. [CrossRef]
38. Sexauer Gustin, M.; Weiss-Penzias, P.S.; Peterson, C. Investigating sources of gaseous oxidized mercury in dry deposition at three sites across Florida, USA. *Atmos. Chem. Phys.* **2012**, *12*, 9201–9219. [CrossRef]
39. Griggs, T.; Liu, L.; Talbot, R.W.; Torres, A.; Lan, X. Comparison of atmospheric mercury speciation at a coastal and an urban site in Southeastern Texas, USA. *Atmosphere* **2020**, *11*, 73. [CrossRef]
40. Sigler, J.M.; Mao, H.; Sive, B.C.; Talbot, R. Oceanic influence on atmospheric mercury at coastal and inland sites: A springtime noreaster in New England. *Atmos Chem Phys* **2009**, *9*, 4023–4030. [CrossRef]

41. Coburn, S.; Dix, B.; Edgerton, E.; Holmes, C.D.; Kinnison, D.; Liang, Q.; ter Schure, A.; Wang, S.; Volkamer, R. Mercury oxidation from bromine chemistry in the free troposphere over the southeastern US. *Atmos. Chem. Phys.* **2016**, *16*, 3743–3760. [CrossRef]
42. Hedgecock, I.M.; Pirrone, N. Chasing Quicksilver: Modeling the Atmospheric Lifetime of Hg in the Marine Boundary Layer at Various Latitudes. *Environ. Sci. Technol.* **2004**, *38*, 69–76. [CrossRef]
43. Yi, J.; Cizdziel, J.; Lu, D. Temporal patterns of atmospheric mercury species in northern Mississippi during 2011–2012: Influence of sudden population swings. *Chemosphere* **2013**, *93*, 1694–1700. [CrossRef]

© 2020 by the authors. Licensee MDPI, Basel, Switzerland. This article is an open access article distributed under the terms and conditions of the Creative Commons Attribution (CC BY) license (http://creativecommons.org/licenses/by/4.0/).

Article

Direct Measurement of Mercury Deposition at Rural and Suburban Sites in Washington State, USA

Marc W. Beutel [1,*], Lanka DeSilva [2,†] and Louis Amegbletor [3]

[1] Department of Civil and Environmental Engineering and Environmental Systems Graduate Group, University of California, Merced, CA 95343, USA
[2] Department of Civil and Environmental Engineering, Washington State University, Pullman, WA 99164, USA; ldesilva@ramboll.com
[3] Environmental Systems Graduate Group, University of California, Merced, CA 95343, USA; lamegbletor@ucmerced.edu
* Correspondence: mbeutel@ucmerced.edu
† Current affiliation: Ramboll, Lynnwood, WA 98036, USA.

Abstract: Because of mercury's (Hg) capacity for long-range transport in the atmosphere, and its tendency to bioaccumulate in aquatic biota, there is a critical need to measure spatial and temporal patterns of Hg atmospheric deposition. Dry deposition of Hg is commonly calculated as the product of a measured atmospheric concentration and an assumed deposition velocity. An alternative is to directly assess Hg deposition via accumulation on surrogate surfaces. Using a direct measurement approach, this study quantified Hg deposition at a rural site (Pullman) and suburban site (Puyallup) in Washington State using simple, low-cost equipment. Dry deposition was measured using an aerodynamic "wet sampler" consisting of a Teflon plate, 35 cm in diameter, holding a thin layer (2.5 mm) of recirculating acidic aqueous receiving solution. In addition, wet Hg deposition was measured using a borosilicate glass funnel with a 20-cm-diameter opening and a 1 L Teflon sampling bottle. Hg deposition was estimated based on changes in total Hg in the aqueous phase of the samplers. Dry Hg deposition was 2.4 ± 1.4 ng/m^2·h (average plus/minus standard deviation; $n = 4$) in Pullman and 1.3 ± 0.3 ng/m^2·h ($n = 6$) in Puyallup. Wet Hg deposition was 7.0 ± 4.8 ng/m^2·h ($n = 4$) in Pullman and 1.1 ± 0.2 ng/m^2·h ($n = 3$) in Puyallup. Relatively high rates of Hg deposition in Pullman were attributed to regional agricultural activities that enhance mercury re-emission and deposition including agricultural harvesting and field burning. Hg concentration in precipitation negatively correlated with precipitation depth, indicating that Hg was scavenged from the atmosphere during the beginning of storm events. Because of their relative simplicity and robustness, direct measurement approaches such as those described in this study are useful in assessing Hg deposition, and for comparing results to less direct estimates and model estimates of Hg deposition.

Keywords: dry deposition; wet deposition; wet sampler; agricultural field burning; AIRPACT

1. Introduction

Mercury (Hg) has a unique ability for long-range transport in the atmosphere [1]. Combined with the tendency of Hg to bioaccumulate in aquatic biota, atmospheric deposition of Hg is a worldwide health concern [2]. Natural sources of Hg are geogenic and include releases from outgassing mantel and crustal materials, volcanoes, and geothermal regions [3]. Key anthropogenic sources of atmospheric Hg include coal combustion, artisanal and small-scale gold mining, metal production, and cement production [4]. Direct anthropogenic sources account for ~30% of total Hg emissions to the atmosphere, natural emissions account for ~10%, and re-emission of previously deposited Hg from oceans and soils make up the remaining ~60% [4,5]. A substantial fraction of deposited Hg ends up in aquatic ecosystems where it can accumulate in aquatic biota as toxic methylmercury [5,6]. As a result of widespread Hg deposition in North America, 6.6 million lake hectares in the U.S. have fish consumption advisories in place due to elevated concentrations of Hg in fish tissue, and 28 states including Washington have statewide fish consumption advisories due mainly to Hg contamination [7]. An important source of Hg deposition in the Pacific Northwest region of the U.S., the geographical focus of this study, is the long-range transport of Hg emissions from Asia [8].

The atmospheric cycling of Hg is complex, but many key processes are known [1–3,9,10]. Atmospheric Hg is dominated by gaseous elemental Hg (GEM), which has a residence time of months, allowing for long-range transport in the atmosphere. GEM can be transformed to gaseous oxidized Hg (GOM) in the atmosphere in the presence of ozone, chlorine gas, hydroxyl radical and bromine compounds. Because of its greater solubility and reactivity GOM has an atmospheric residence time of days to weeks. GOM, with its relatively low vapor pressure, can partition onto particulate matter to form particulate-bound Hg (PBM), which, like GOM, has a relatively short residence time that is dependent on particle size. The surface flux of Hg is bidirectional since deposited Hg can efflux back into the atmosphere as GEM. Hg deposition can occur as dry or wet deposition. Depending on season and setting, dry deposition can be dominated by the deposition of gaseous Hg species or by the deposition of PBM. The source of Hg in wet deposition includes in-cloud oxidation of GEM to soluble Hg(II) and scavenging of GOM and PBM out of the atmosphere.

Rates of dry deposition are commonly calculated as the product of a measured Hg atmospheric concentration and an assumed deposition velocity [11]. The magnitude of deposition velocity is relatively small for GEM (0.1–0.4 cm/s), moderate for PBM (0.02–2 cm/s), and large for GOM (0.5–6 cm/s). An alternative method to estimate Hg dry deposition is a direct approach that captures deposited Hg onto a variety of abiotic surfaces including water surfaces, membranes and artificial turf [12–15]. This method has the advantage of obviating the need to assume a deposition velocity. However, uncertainty exists as to how well surrogate surfaces mimic environmental surfaces. In this study we used a direct wet-sampler to measure dry deposition in the summer and fall of 2011 to compare patterns at a rural site in eastern Washington and a suburban site in western Washington. Wet deposition was also assessed with a simple funnel and Teflon sampling bottle. Since urban locations tend to have higher rates of Hg deposition, we anticipated that the urban site would exhibit higher rates of Hg deposition. However, as illustrated in this study and acknowledged in other studies, this linkage can be weak since Hg deposition is affected not only by local emissions but also atmospheric processes [1,16]. We also assessed the significance of ephemeral summer storms on the annual Hg deposition budget for the rural site in eastern Washington and compared directly measured Hg deposition to results from a regional air quality model (AIRPACT-3).

This project was motivated by calls for expanded efforts to complement indirect measurement of Hg deposition with direct measurements that assess both dry and wet Hg deposition using inexpensive and simple approaches [14,17–19]. It was also motivated by the limited amount of Hg deposition data in Washington State and the greater Pacific Northwest [20–22]. While the data presented here is from a study several years ago, we were prompted to publish it because of its good fit with this special issue of Atmosphere focusing on atmospheric mercury monitoring. We believe our study's results are still relevant and highlight the utility of direct monitoring of mercury deposition.

2. Experiments

2.1. Sampling Sites

Hg deposition was measured at two contrasting sites in Washington State (Figure 1). One site was located in Pullman in rural eastern Washington (46.733° N, 117.172° W). Samples were collected on the rooftop of Dana Hall, a three-story building (9.8 m height) on the Washington State University (WSU) campus. Pullman is surrounded by wheat, barley, lentil, and pea agricultural fields. The regional climate is semi-arid with hot, dry summers and cold, wet winters. Annual precipitation is ~50 cm/y. Of potential importance to this study, agricultural field burning in the region is common in the fall, mainly during August and September. Dry deposition samples were collected in Pullman during four precipitation-free sampling events in August–September 2011. Wet deposition was measured during four events in June, July and October 2011. Rainfall events ranged in duration from 2 to 21 h and had mean precipitation intensity of 0.08–0.61 mm/h.

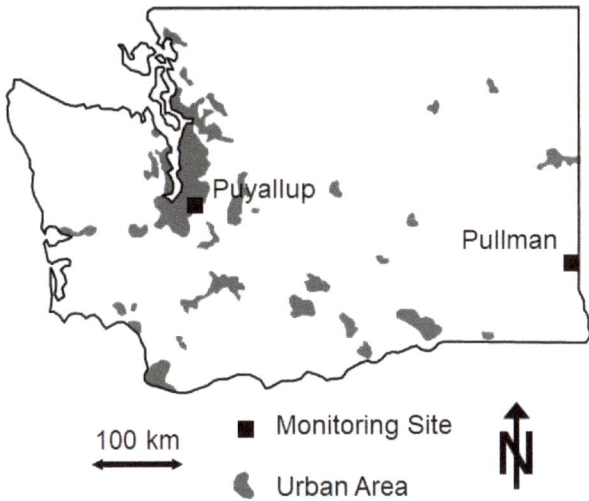

Figure 1. Site locations of deposition monitoring.

The other site was located in suburban Puyallup in western Washington (47.185° N, 122.292° W). Samples were collected on the rooftop of the Avian Health and Food Safety Laboratory (4 m height) at the WSU Puyallup Research and Extension Center. The site is southeast of the heavily urbanized and industrialized Seattle–Tacoma metropolitan area. The Centralia Power Plant, which is 73 km southwest of the study site, is the only coal fire power plant in Washington and is a major source of Hg deposition to western Washington. Puyallup is also relatively vulnerable to Hg deposition from long-range atmospheric transport of Hg from Asia [9]. The regional climate of Puyallup is cool with dry summers and mild, wet, and cloudy winters. Annual precipitation is ~100 cm/y. Dry deposition samples were collected during six sampling events in September–October 2011. Wet deposition was measured during three events in September–October 2011. Rainfall events ranged in duration from 16 to 56 h and had mean precipitation intensity of 0.30–0.38 mm/h.

2.2. Dry Deposition Sampling

The dry deposition sampling apparatus was based on [23,24]. A wet sampler was used that collected dry deposition in an acidic aqueous solution that was analyzed for total Hg (THg). The wet sampler was anticipated to mainly collect deposited GOM and PBM, but also some GEM since the solubility of elemental Hg is enhanced in low-pH water [25]. The sampler consisted of a Teflon plate, 35 cm in diameter, holding a thin layer (2.5 mm) of aqueous receiving solution (Figure 2). The plate was held by a collector with an outer edge shaped like an airfoil to minimize air flow disturbance over the water surface. The receiving solution was continuously circulated onto the plate via a pump and Teflon tubing. Solution was discharge onto the top of the plate then flowed over four weirs on the outer edge of the plate and into the collector, which was connected to a 5 L glass reservoir. The reservoir was chilled by a refrigeration unit to 10 °C to minimize evaporative losses.

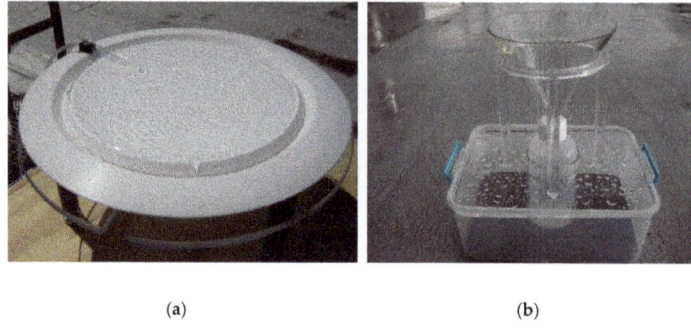

(a) (b)

Figure 2. (**a**): Dry deposition sampler. Plate diameter is 35 cm. (**b**): Wet deposition sampler. Funnel diameter is 20 cm.

Before each sampling event, the water surface holder, plate, tubing, and reservoir were cleaned thoroughly with 25% nitric acid and triple rinsed with reagent grade deionized water. The reservoir was filled with 5 L of 0.05 M hydrochloric acid (HCl) receiving solution which was circulated through the system at a rate of 200 mL/min. After 30 min of operation, an initial sample of receiving solution was collected for later Hg analysis. The system was operated for several days and a second sample of receiving solution was collected at the end of the sampling event. To determine the final volume, the remaining receiving solution was decanted and weighed. In the first Pullman and all Puyallup sampling events, two samples were collected, one mid-way through the sampling event and another at the end of the sampling event.

Samples were preserved with 1% bromine monochloride and analyzed in triplicate for THg based on United States Environmental Protection Agency (USEPA) Method 1631 [26] on a Brooks Rand MERX-T mercury auto analyzer. The method detection limit for THg is 0.2 ng/L. Standard quality control procedures for THg analyses included calibration blanks (acceptable range < 50 pg; mean = 2.98 pg, n = 20), matrix spike samples (acceptable range 71–125%; mean recovery = 88.6%, n = 15), and ongoing precision (acceptable range 77–123%; mean recovery = 96.6%, n = 24). Mass accumulation of THg was estimated as the difference between the concentration multiplied by associated volume of receiving solution at the end and beginning of the sampling event. Dry Hg deposition in ng/m^2·hr was calculated as the mass accumulation divided by the area of the plate (0.0962 m^2) and the duration of the sampling event.

2.3. Wet Deposition Sampling

Wet deposition sampling apparatus was based on [27]. The sampler included four components: a borosilicate glass funnel with a 20-cm-diameter opening, a Teflon adaptor, a 1 L Teflon sampling bottle, and a Plexiglas holder (Figure 2). Prior to each sampling event, the borosilicate glass funnel, adaptor, and sampling bottle were cleaned thoroughly with 25% nitric acid and triple rinsed with reagent grade deionized water. Twenty ml of 0.08 M HCl receiving solution was added to the sample bottle to enhance capture and preservation of Hg in collected precipitation. A subsample of receiving solution was collected at the beginning of the sampling event for later THg analysis. At the end of the sampling event, collected rainwater was weighed to determine the volume of precipitation, then a sample was collected. Samples were preserved and analyzed for THg as described above for the dry deposition monitoring, with the exception that Pullman June and July samples were not run in triplicate due to limited sample volume. Mass accumulation of Hg was estimated as the THg concentration in the rainfall-receiving solution mixture collected at the end of the sampling event multiplied by its volume, corrected for the initial THg mass in the receiving solution. Wet Hg deposition in ng/m^2·hr was calculated as mass accumulation divided by the area of the funnel (0.0314 m^2) and the duration of the sampling event.

3. Results

3.1. Dry Deposition

Dry Hg deposition measured in Pullman ranged from 1.0 to 4.3 ng/m²·h and averaged 2.4 ng/m²·h (Table 1, Figure 3). The high deposition rate observed in late August (4.3 ng/m²·h) corresponded with smoky conditions in Pullman. High air temperatures (12–26 °C), low relative humidity (31–54%), and elevated wind speeds (0.8–3.1 m/s) during August sampling resulted in high rates of evaporation [28]. Normalized to the area of the sampler, evaporation rates (5.4–5.9 mm/d) were similar to the historical mean August pan evaporation rate for the area (6.7 mm/d) [29]. Dry Hg deposition in Puyallup ranged from 0.8 to 1.6 ng/m²·h and averaged 1.3 ng/m²·h. Relative to Pullman, dry deposition in Puyallup was lower in magnitude and less variable. Measured evaporation rates were relatively low and corresponded with low air temperature (11–22 °C), high relative humidity (66–88%), and low wind speed (0–0.6 m/s) during September sampling [28]. Estimated evaporation rates for the three events (0.3–2.0 mm/d) were lower than the historical mean September pan evaporation rate for the region (2.4 mm/d) [29].

Table 1. Mercury dry deposition.

Date	Duration (Days)	Initial Vol (L)	Final Vol (L)	Evap Rate (L/d)	Initial Total Hg [a] (ng/L)	Final Total Hg [a] (ng/L)	Hg Dry Dep [b] (ng/m²·h)
Pullman, WA							
15–18 Aug	2.77	4.87	3.28 [c]	0.57 [d]	1.57	7.12	2.5
18–22 Aug	3.96	3.02	0.75		7.12	41.1	1.0
23–28 Aug	5.04	4.74	1.96	0.55	1.43	28.9	4.3
29 Aug–5 Sep	7.11	4.74	1.02	0.52	1.26	30.8	1.6
Puyallup, WA							
12–15 Sep	3.03	4.75	4.26 [c]	0.16 [d]	1.37	4.01	1.5
15–17 Sep	1.64	4.01	3.74		4.01	5.14	0.8
21–23 Sep	2.05	4.75	4.69 [c]	0.03 [d]	0.79	2.37	1.6
23–25 Sep	1.74	4.43	4.38		2.37	3.45	1.2
27–29 Sep	2.00	4.75	4.37 [c]	0.19 [d]	0.68	2.21	1.4
29 Sep–2 Oct	2.97	4.11	3.55		2.21	4.48	1.0

[a] Average of triplicate analysis. [b] Deposition based on sampler cross-sectional area of 0.096 m². [c] Volume estimated based on evaporation rate over entire event. [d] Evaporation rate based on total volume loss over entire event.

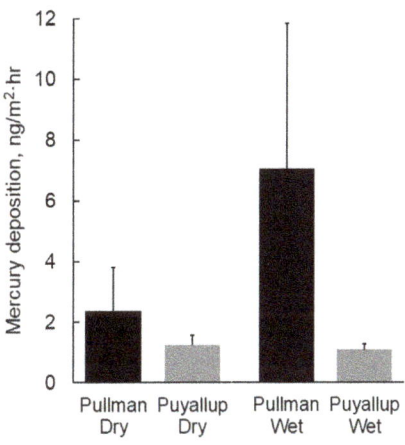

Figure 3. Dry and wet Hg deposition at Pullman and Puyallup, WA. Values are mean plus one standard deviation (n = 3–6).

3.2. Wet Deposition

Wet Hg deposition measured in Pullman ranged from 1.8 to 12.8 ng/m^2·h and averaged 7.0 ng/m^2·d (Table 2, Figure 3). Deposition rates for the three storm events in June/July were an order of magnitude higher than for the October event. Two summer events in June and July with low precipitation intensity (<0.17 mm/h) had relatively high mean Hg concentrations in precipitation (>56 ng/L). Wet Hg deposition in Puyallup in September/October ranged from 0.9 to 1.3 ng/m^2·h. Mean Hg concentrations in Puyallup precipitation ranged from 2.3 to 3.5 ng/L. As with dry deposition, wet deposition rates and Hg concentrations in precipitation were lower and less variable in Puyallup when compared to Pullman.

Table 2. Mercury wet deposition.

Date	Duration (h)	Initial Vol (mL)	Final Vol (mL)	Precip (mm/h)	Initial Total Hg [a] (ng/L)	Final Total Hg [a] (ng/L)	Hg Wet Dep [c] (ng/m^2·h)	Precip Hg (ng/L)
Pullman, WA								
28–29 June	3.50	20	38.9	0.17	0.81	36.1 [b]	12.8	73.4
12–13 July	11.0	20	48.5	0.08	0.90	33.8 [b]	4.8	56.9
14 July	2.17	20	61.2	0.61	0.88	9.84 [b]	8.7	14.2
10–11 Oct	21.0	20	210	0.29	0.27	5.74	1.8	6.32
Puyallup, WA								
17–19 Sep	37.9	20	467	0.38	0.71	2.27	0.89	2.34
25–27 Sep	55.4	20	533	0.30	0.86	3.28	1.0	3.37
2–3 Oct	16.1	20	209	0.38	0.62	3.18	1.3	3.45

[a] Average of triplicate analysis. [b] Single analysis due to low precipitation volume. [c] Deposition based on funnel cross sectional area of 0.031 m^2.

4. Discussion

4.1. Dry Deposition

Dry Hg deposition rates observed in this study, typically 1–2 ng/m^2·h, were similar to a limited number of studies that report direct measurements of dry Hg deposition using wet samplers with acidified de-ionized water as receiving solution. Dry Hg deposition measured in Komae, a heavily industrialized city in the western Tokyo metropolitan area, ranged from 0.5 to 3 ng/m^2·h (n = 5) [24]. Related long-term studies measured Hg deposition monthly at 10 sites in Japan over multiple years. Mean dry deposition ranged from 0.4–2 ng/m^2·h and higher deposition rates were associated with urban and industrial areas [30,31]. Dry deposition ranging from 0.4 to 1.7 ng/m^2·h (n = 5) was measured in rural New York State [32]. Some studies have used un-acidified de-ionized water as receiving water in passive static samplers and report dry deposition rates ranging from 0.5 to 2 ng/m^2·h in an urban site in Ohio [14] and 0.3–0.5 ng/m^2·h in the Florida Everglades [33]. Measured deposition rates generally increase with the addition of acid presumably due to enhanced capture and retention of GEM [25].

In an effort to compare our direct measurements of dry deposition with those estimated based on atmospheric Hg monitoring, we searched the USEPA AirData Air Quality Monitors website and found relevant data in the National Air Toxics Trends Stations (NATTS) database [34]. While no stations were located in the rural eastern portion of Washington, a station was located near our Puyallup site (site code 53/033/0080; 47.56824° N 122.309° W). Reported arithmetic mean daily values for Hg PM10 concentrations for 2011 ranged from 0 to 0.04 ng/m^3 over the year (n = 60) and were ~0.01 ng/m^3 during this study (Table S1). Back calculating a depositional velocity during the study period as the mean measured dry deposition flux (1.25 ng/m^2·h; Table 1) divided by the mean measured Hg PM10 (0.0075 ng/m^3; Table S1) yields a deposition velocity on the order of 2.6 cm/s which is similar in magnitude but on the high end of deposition velocity typically reported for PBM (0.02–2 cm/s) [11].

The rural site in Pullman exhibited rates of dry Hg deposition that were somewhat higher in magnitude when compared to the suburban Puyallup site. While rural, the landscape around Pullman is intensely managed for agricultural production. When crops are harvested, Hg deposited on soil and plants during the growing season can be dispersed into the atmosphere on dust and deposited back onto the landscape, resulting in enhanced localized Hg deposition [35]. Agricultural field tilling has also been implicated in elevated levels of atmospheric Hg and downwind Hg deposition [36]. Wildfires and agricultural fires can also substantially enhance Hg emissions via combustion of biomass, litter and organic soils, with the magnitude of emissions apparently corresponding with fire severity [10]. A 3–4 fold increase in total gaseous Hg was measured above wild and agricultural fires in eastern Washington and Oregon [37]. Hg emitted via fires can enhance regional Hg deposition rates [3]. A 12-fold increase in PBM deposition was measured in New Mexico resulting from transport of smoke from a large forest fire in Arizona [38].

A review of regional burn permits, along with regional meteorology and air quality, suggests that the relatively elevated dry Hg deposition in Pullman measured between 15 August and 5 September was affected by agricultural field burning. Burn permits allotted by the Washington State Department of Ecology indicated that agricultural field burns totaling 400 ha occurred in Walla Walla County, 160 km southwest of Pullman, between 15 and 27 August. Prevailing wind direction in the region during August 2011 was generally southwesterly. Mean hourly wind direction from 15 to 27 August was $236° \pm 68°$ (mean $\pm$ standard deviation, $n = 142$) in Pullman (http://mesowest.utah.edu/; station KPUW) and $183° \pm 70°$ ($n = 266$) in Walla Walla (http://mesowest.utah.edu/; station KALW). Back trajectory modeling using HYSPLIT, a hybrid single-particle lagrangian integrated trajectory model hosted by the Nation Oceanic and Atmospheric Administration (http://www.arl.noaa.gov/HYSPLIT.php), confirmed that air parcels in Pullman generally passed through the Walla Walla region throughout mid to late August 2011 (Figure S1). In addition, because wildfires and prescribed fires are known to be substantial sources of atmospheric particulate matter [39], we assessed PM2.5 concentrations reported via the USEPA Outdoor Air Quality Data repository (https://www.epa.gov/outdoor-air-quality-data/download-daily-data) for Pullman (station 530710005) and Walla Walla (station 530750003). PM2.5 concentrations were generally <5 µg/m^3 in July and <5 µg/m^3 in August, with levels increasing to >10 µg/m^3 at times in late August and early September (Figure S2), suggesting that smoky conditions were present in Pullman when dry deposition measurements were performed in this study. Relatively high levels of PM2.5 in Pullman between 23 and 28 August (Figure S2) corresponded with the highest measured Hg dry deposition of 4.3 ng/m^2·h (Table 1), further suggesting a linkage between smoky conditions and Hg deposition.

4.2. Wet Deposition

Except for the June and July events in Pullman, which had elevated THg concentrations in precipitation of 73 ng/L and 57 ng/L, respectively, the results of this study were comparable to concentrations reported worldwide, which typically range from 1 to 15 ng/L [40]. Concentrations are also similar to those measured in California precipitation, which ranged from 1 to 28 ng/L and averaged 4 ng/L ($n = 46$) [41]. The extreme concentrations measured in Pullman appear to be relatively uncommon at other reported monitoring sites. In hundreds of precipitation samples from Europe and China, 95th percentile THg concentrations rarely exceeded 40 ng/L [40]. The high Pullman THg concentrations were undoubtedly related to the small precipitation amount associate with these monitored events. While annual Hg deposition generally correlates with total annual precipitation, THg concentration in collected precipitation generally shows a negative correlation with precipitation amount [1]. This observation is attributed to scavenging of Hg from the air column early during storm events and subsequent dilution of the accumulating sample later in the event. The relationship between THg concentration and precipitation amount was apparent in our pooled samples (Figure 4). This phenomenon, combined with the fact that agricultural activities can enhance Hg re-emission from the landscape [35,36] explains the high Hg concentrations in precipitation observed during small rain events in Pullman in June and July.

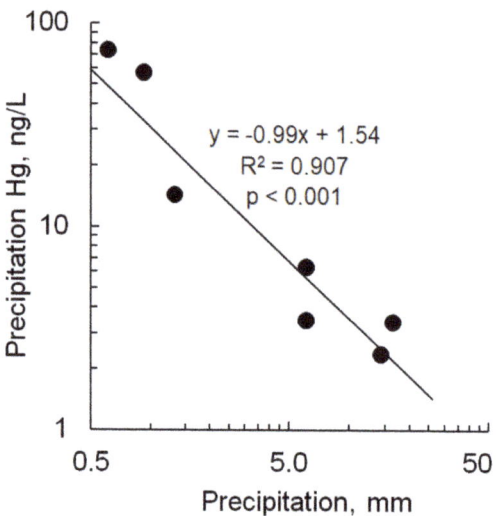

Figure 4. Relationship between Hg concentration in precipitation and precipitation depth for data for Pullman and Puyallup, WA. Line is linear regression.

While brief in duration, wet Hg deposition via ephemeral storm events observed in Pullman in June and July deposited a similar magnitude of Hg to longer storm events. For example, the 3.5-h-long June storm event deposited 45 ng/m^2 while the 21-h-long October storm event deposited 39 ng/m^2. While the frequency of monitoring in this study was admittedly limited, the data obtained can be used to assess patterns of Hg deposition during the dry period in Pullman (May–September 2011). Based on an examination of 2011 precipitation patterns for the dry period in Pullman and mass balance considerations, ephemeral storm events may have accounted for ~ 20% of dry-season Hg deposition (see Section S1 in Supplemental Material for additional details).

4.3. Model Versus Direct Measurements of Hg Deposition

AIRPACT-3 (Air Indicator Report for Public Access and Community Tracking) is a numerical air quality forecast system that operates daily for the Pacific Northwest. It estimates a range of air quality parameters including Hg dry and wet deposition (see Section S2 in Supplemental Material for additional details). The modeled rates of Hg deposition from AIRPACT-3 were generally lower, but of similar magnitude, compared to direct measurements in the field. For Pullman (Figure S3), modeled dry deposition typically ranged from 1 to 5 ng/m^2·d with occasional peaks of 20–40 ng/m^2·d. Modeled peaks of dry Hg deposition were similar in magnitude but shorter in duration than direct measurements in mid-August (37 ng/m^2·d) and early September (39 ng/m^2·d). Modeled values were far below the extreme dry deposition rate of 103 ng/m^2·d measured in late August, likely associated with enhanced deposition from regional agricultural field burning. Modeled rates of wet deposition showed four events with substantial deposition (>5 ng/m^2·d), one in June, one in July, and two in October. The July and October modeled deposition events coincided temporally with direct measurement field events, indicating that the model captured actual precipitation events, including occasional ephemeral storm events during the summer. However, direct measurements on those two dates were 5–10 times the model estimates.

For Puyallup, modeled dry deposition typically ranged from 3 to 15 ng/m^2·d with occasional peaks of 30–60 ng/m^2·d (Figure S4). Measured dry deposition was higher, ranging from 28 to 33 ng/m^2·d. The model captured the temporal dynamics of precipitation events associated with direct field measurements fairly well, including dry conditions during all three dry deposition monitoring events and wet conditions during the second and third wet deposition monitoring events, though the timing of the third event in early October was a bit early relative to observed precipitation in Puyallup. Based on the last two monitoring events in late September and early October, direct measurements of wet deposition were 2–4 times higher than those predicted by the model. Considering that this study compared a limited number of direct measurements at discrete points in space to modeling results on a 12-km grid, rates of Hg deposition predicted by AIRPACT-3 were reasonably close to direct measurements.

5. Conclusions

This study used a direct approach to assess the magnitude of dry and wet Hg deposition at a rural and suburban site in Washington. Measured values, generally ranging from 1 to 2 ng/m^2·h, are some of the first published direct measurements of Hg deposition in the State. Key conclusions of the study include:

(1) Because of their relative simplicity and robustness, direct measurement approaches such as those described in this study are useful in assessing temporal and spatial patterns of Hg deposition, and for comparing results to less direct estimates of Hg deposition and estimates from numerical air quality models.

(2) Hg deposition can be substantial in rural regions with significant agricultural activities. Hg deposition rates at the rural study site (Pullman, Washington) were similar to or higher than deposition rates observed at the suburban study site (Puyallup, Washington), which was likely influenced by regional urban and industrial sources of Hg.

(3) In rural agricultural areas, agricultural burning and associated re-emission and transport of previously deposited Hg can lead to elevated levels of Hg dry deposition. Rates of dry deposition in Pullman during smoky conditions indicative of agricultural burning were ~2.5 times the deposition rates observed during non-smoky conditions.

(4) Hg concentrations in precipitation correlated negatively with precipitation depth, suggesting that scavenging of PBM and GOM from the atmosphere at the beginning of storm events was an important wet deposition process.

(5) Ephemeral, short-term storm events at the rural Pullman site had elevated Hg concentrations. Mass balance estimates indicated that these Hg-rich storm events may account for a meaningful fraction (~20%) of dry season Hg deposition.

Supplementary Materials: The following are available online at https://www.mdpi.com/2073-4433/12/1/35/s1: Table S1: Atmospheric Hg PM10 concentrations measured near Puyallup, WA; Figure S1: Example HYSPLIT back trajectories from Pullman, WA; Figure S2: PM2.5 data for Pullman and Walla Walla, WA; Section S1. Ephemeral Storm Events as Source of Hg Deposition; Section S2. AIRPACT-3 Air Quality Model; Figure S3: Measured and modeled dry and wet Hg deposition for Pullman, WA; Figure S4: Measured and modeled dry and wet Hg deposition for Puyallup, WA.

Author Contributions: Conceptualization, methodology, formal analysis, M.W.B. and L.D.; investigation, M.W.B., L.D. and L.A.; writing—original draft preparation, M.W.B. and L.D.; writing—review and editing, M.W.B. All authors have read and agreed to the published version of the manuscript.

Funding: This project was funded in part by the National Science Foundation (#0846446) and the Atmospheric Policy Trajectory Program of the Washington State University Laboratory for Atmospheric Research.

Institutional Review Board Statement: Not applicable.

Informed Consent Statement: Not applicable.

Data Availability Statement: Data collected in this study is, for the most part, presented in the paper and Supplemental Material. Any additional data is available upon request from the corresponding author.

Acknowledgments: This project was completed when Beutel was a faculty member and Lanka DaSilva was in the Environmental Engineering Master's degree program in the Civil and Environmental Engineering Department at Washington State University. We would like to thank the following people for their assistance during this project: Brian Lamb and Joseph Vaughn from Laboratory for Atmospheric Research at Washington State University, particularly for suppling HYSPLIT and AIRPACT-3 modeling results; John Stark and the staff of the Washington State University Puyallup Research and Extension Center; and Philip Kenyon, Washington State University undergraduate research assistant. We would also like to thank the anonymous reviewers for their constructive comments on the manuscript. The views expressed herein are solely those of the authors and do not represent the official policies or positions of any supporting agencies.

Conflicts of Interest: The authors declare no conflict of interest.

References

1. Lyman, S.N.; Cheng, I.; Gratz, L.E.; Weiss-Penzias, P.; Zhang, L. An updated review of atmospheric mercury. *Sci. Total Environ.* **2020**, *707*, 135575. [CrossRef] [PubMed]
2. Driscoll, C.T.; Mason, R.P.; Chan, H.M.; Jacob, D.J.; Pirrone, N. Mercury as a Global Pollutant: Sources, Pathways, and Effects. *Environ. Sci. Technol.* **2013**, *47*, 4967–4983. [CrossRef] [PubMed]
3. Schroeder, W.H.; Munthe, J. Atmospheric mercury—an overview. *Atmos. Environ.* **1998**, *32*, 809–822.
4. United Nations Environment Programme (UNEP). *Global Mercury Assessment 2019: Sources, Emissions, Releases and Environmental Transport*; UNEP Chemicals Branch: Geneva, Switzerland, 2019.
5. Lindberg, S.E.; Bullock, R.O.; Ebinghaus, R.; Engstrom, D.R.; Feng, X.; Fitzgerald, W.F.; Pirrone, N.; Prestbo, E.M.; Seigneur, C. A synthesis of progress and uncertainties in attributing the sources of mercury in deposition: Panel on source attribution of atmospheric mercury. *Ambio* **2007**, *36*, 19–32.
6. Swain, E.; Jakus, P.; Rice, G.; Lupi, F.; Maxson, P.; Pacyna, J.; Penn, A.; Spiegel, S.; Veiga, M. Socioeconomic consequences of mercury use and pollution. *Ambio* **2007**, *36*, 45–61.
7. United State Environmental Protection Agency (USEPA). *Biennial National Listing of Fish Advisory*; EPA-820-F-11-014; USEPA: Washington, DC, USA, 2011.
8. Jaffe, D.; Prestbo, E.; Swartzendruber, P.; Weisspenzias, P.; Kato, S.; Takami, A.; Hatakeyama, S.; Kajii, Y. Export of atmospheric mercury from Asia. *Atmos. Environ.* **2005**, *39*, 3029–3038. [CrossRef]
9. Lin, C.-J.; Pehkonen, S.O. The chemistry of atmospheric mercury: A review. *Atmos. Environ.* **1999**, *33*, 2067–2079. [CrossRef]
10. Obrist, D.; Kirk, J.L.; Zhang, L.; Sunderland, E.M.; Jiskra, M.; Selin, N.E. A review of global environmental mercury processes in response to human and natural perturbations: Changes of emissions, climate, and land use. *Ambio* **2018**, *47*, 116–140. [CrossRef]
11. Zhang, L.; Wright, L.P.; Blanchard, P. A review of current knowledge concerning dry deposition of atmospheric mercury. *Atmos. Environ.* **2009**, *43*, 5853–5864. [CrossRef]
12. Hall, N.L.; Dvonch, J.T.; Marsik, F.J.; Barres, J.A.; Landis, M. An Artificial Turf-Based Surrogate Surface Collector for the Direct Measurement of Atmospheric Mercury Dry Deposition. *Int. J. Environ. Res. Public Health* **2017**, *14*, 173. [CrossRef]
13. Huang, J.; Lyman, S.N.; Hartman, J.S.; Gustin, M.S. A review of passive sampling systems for ambient air mercury measurements. *Environ. Sci. Process. Impacts* **2014**, *16*, 374–392. [CrossRef] [PubMed]
14. Huang, J.; Choi, H.-D.; Landis, M.S.; Holsen, T.M. An application of passive samplers to understand atmospheric mercury concentration and dry deposition spatial distributions. *J. Environ. Monit.* **2012**, *14*, 2976–2982. [CrossRef] [PubMed]
15. Lyman, S.N.; Gustin, M.S.; Prestbo, E.M.; Kilner, P.I.; Edgerton, E.; Hartsell, B. Testing and Application of Surrogate Surfaces for Understanding Potential Gaseous Oxidized Mercury Dry Deposition. *Environ. Sci. Technol.* **2009**, *43*, 6235–6241. [CrossRef] [PubMed]
16. Sprovieri, F.; Pirrone, N.; Ebinghaus, R.; Kock, H.; Dommergue, A. A review of worldwide atmospheric mercury measurements. *Atmos. Chem. Phys. Discuss.* **2010**, *10*, 8245–8265. [CrossRef]
17. Gustin, M.S.; Jaffe, D. Reducing the Uncertainty in Measurement and Understanding of Mercury in the Atmosphere. *Environ. Sci. Technol.* **2010**, *44*, 2222–2227. [CrossRef]
18. Lyman, S.N.; Gustin, M.S.; Prestbo, E.M.; Marsik, F.J. Estimation of Dry Deposition of Atmospheric Mercury in Nevada by Direct and Indirect Methods. *Environ. Sci. Technol.* **2007**, *41*, 1970–1976. [CrossRef]
19. Zhang, L.; Lyman, S.; Mao, H.; Lin, C.-J.; Gay, D.A.; Wang, S.; Gustin, M.S.; Feng, X.; Wania, F. A synthesis of research needs for improving the understanding of atmospheric mercury cycling. *Atmos. Chem. Phys. Discuss.* **2017**, *17*, 9133–9144. [CrossRef]
20. Prestbo, E.M.; Gay, D. Wet deposition of mercury in the U.S. and Canada, 1996–2005: Results and analysis of the NADP mercury deposition network (MDN). *Atmos. Environ.* **2009**, *43*, 4223–4233. [CrossRef]

1. Zhang, L.; Wu, Z.; Cheng, I.; Wright, L.P.; Olson, M.L.; Gay, D.A.; Risch, M.R.; Brooks, S.; Castro, M.S.; Conley, G.D.; et al. The Estimated Six-Year Mercury Dry Deposition Across North America. *Environ. Sci. Technol.* **2016**, *50*, 12864–12873. [CrossRef]
2. Era-Miller, B. *Toxics Atmospheric Deposition in Eastern Washington State–Literature Review*; EA Project Code 10-124; Washington State Department of Ecology: Olympia, WA, USA, 2011.
3. Yi, S.-M.; Holsen, T.M.; Noll, K.E. Comparison of Dry Deposition Predicted from Models and Measured with a Water Surface Sampler. *Environ. Sci. Technol.* **1997**, *31*, 272–278. [CrossRef]
4. Sakata, M.; Marumoto, K. Dry Deposition Fluxes and Deposition Velocities of Trace Metals in the Tokyo Metropolitan Area Measured with a Water Surface Sampler. *Environ. Sci. Technol.* **2004**, *38*, 2190–2197. [CrossRef] [PubMed]
5. Waite, D.T.; Snihura, A.D.; Liu, Y.; Huang, G. Uptake of atmospheric mercury by deionized water and aqueous solutions of inorganic salts at acidic, neutral and alkaline pH. *Chemosphere* **2002**, *49*, 341–351. [CrossRef]
6. United State Environmental Protection Agency (USEPA). *Method 1631, Revision E: Mercury in Water by Oxidation, Purge and Trap, and Cold Vapor Atomic Fluorescence Spectrometry*; EPA-821-R-02-019; USEPA: Washington, DC, USA, 2002.
7. Landis, M.S.; Keeler, G.J. Critical Evaluation of a Modified Automatic Wet-Only Precipitation Collector for Mercury and Trace Element Determinations. *Environ. Sci. Technol.* **1997**, *31*, 2610–2615. [CrossRef]
8. MesoWest Data. Applicable mean daily values accessed for Pullman station KPUW and Puyallup station PLU. Available online: mesowest.utah.edu/ (accessed on 1 March 2013).
9. Western Regional Climate Center. Data reported for Pullman measured in nearby Moscow, Idaho and for Puyallup measured at Puyallup 2 West Experimental Station. Available online: www.wrcc.dri.edu/ (accessed on 1 March 2013).
10. Sakata, M.; Asakura, K. Evaluating Relative Contribution of Atmospheric Mercury Species to Mercury Dry Deposition in Japan. *WaterAirSoil Pollut.* **2008**, *193*, 51–63. [CrossRef]
11. Sakata, M.; Marumoto, K.; Narukawa, M.; Asakura, K. Regional variations in wet and dry deposition fluxes of trace elements in Japan. *Atmos. Environ.* **2006**, *40*, 521–531. [CrossRef]
12. Lai, S.-O.; Huang, J.; Hopke, P.K.; Holsen, T.M. An evaluation of direct measurement techniques for mercury dry deposition. *Sci. Total. Environ.* **2011**, *409*, 1320–1327. [CrossRef]
13. Marsik, F.J.; Keeler, G.J.; Landis, M.S. The dry-deposition of speciated mercury to the Florida Everglades: Measurements and modeling. *Atmos. Environ.* **2007**, *41*, 136–149. [CrossRef]
14. USEPA AirData Air Quality Monitors. Available online: https://epa.maps.arcgis.com/apps/webappviewer/index.html?id=5f239fd3e72f424f98ef3d5def547eb5&extent=-146.2334,13.1913,-46.3896,56.5319 (accessed on 1 December 2020).
15. Cobbett, F.; Vanheyst, B. Measurements of GEM fluxes and atmospheric mercury concentrations (GEM, RGM and Hgp) from an agricultural field amended with biosolids in Southern Ont., Canada (October 2004–November 2004). *Atmos. Environ.* **2007**, *41*, 2270–2282. [CrossRef]
16. Bash, J.O.; Miller, D.R. A note on elevated total gaseous mercury concentrations downwind from an agriculture field during tilling. *Sci. Total. Environ.* **2007**, *388*, 379–388. [CrossRef]
17. Friedli, H.R.; Radke, L.F.; Prescott, R.; Hobbs, P.V.; Sinha, P. Mercury emissions from the August 2001 wildfires in Washington State and an agricultural waste fire in Oregon and atmospheric mercury budget estimates. *Glob. Biogeochem. Cycles* **2003**, *17*, 1039. [CrossRef]
18. Caldwell, C.A.; Swartzendruber, P.; Prestbo, E. Concentration and Dry Deposition of Mercury Species in Arid South Central New Mexico (2001–2002). *Environ. Sci. Technol.* **2006**, *40*, 7535–7540. [CrossRef]
19. O'Dell, K.; Ford, B.; Fischer, E.V.; Pierce, J.R. Contribution of Wildland-Fire Smoke to US PM2.5 and Its Influence on Recent Trends. *Environ. Sci. Technol.* **2019**, *53*, 1797–1804. [CrossRef]
20. Sprovieri, F.; Pirrone, N.; Bencardino, M.; D'Amore, F.; Angot, H.; Barbante, C.; Brunke, E.-G.; Arcega-Cabrera, F.; Cairns, W.; Comero, S.; et al. Five-year records of mercury wet deposition flux at GMOS sites in the Northern and Southern hemispheres. *Atmos. Chem. Phys. Discuss.* **2017**, *17*, 2689–2708. [CrossRef]
21. Gill, G. Task 3—Atmospheric Mercury Deposition Studies. In *Transport, Cycling, and Fate of Mercury and Monomethyl Mercury in the San Francisco Delta and Tributaries: An Integrated Mass Balance Assessment Approach*; Calfed Mercury Project 2008 Report; Moss Landing Marine Laboratory: Moss Landing, CA, USA, 2008.

Article

Atmospheric Mercury Deposition in Macedonia from 2002 to 2015 Determined Using the Moss Biomonitoring Technique

Trajče Stafilov [1,*], Lambe Barandovski [2,*], Robert Šajn [3] and Katerina Bačeva Andonovska [4]

1. Institute of Chemistry, Faculty of Natural Sciences and Mathematics, Ss. Cyril and Methodius University, POB 162, 1000 Skopje, Macedonia
2. Institute of Physics, Faculty of Natural Sciences and Mathematics, Ss. Cyril and Methodius University, POB 162, 1000 Skopje, Macedonia
3. Geological Survey of Slovenia, Dimičeva 14, 1000 Ljubljana, Slovenia; Robert.Sajn@GEO-ZS.SI
4. Research Center for Environment and Materials, Academy of Sciences and Arts of the Republic of North Macedonia—MANU, Krste Misirkov 2, 1000 Skopje, Macedonia; kbaceva@manu.edu.mk
* Correspondence: trajcest@pmf.ukim.mk (T.S.); lambe@pmf.ukim.mk (L.B.); Tel.: +38970350756 (T.S.); +38970607921 (L.B.)

Received: 2 November 2020; Accepted: 18 December 2020; Published: 21 December 2020

Abstract: The moss biomonitoring technique was used in 2002, 2005, 2010 and 2015 in a potentially toxic elements study (PTEs) in Macedonia. For that purpose, more than 70 moss samples from two dominant species (*Hypnum cupressiforme* and *Homalothecium lutescens*) were collected during the summers of the mentioned years. Total digestion of the samples was done using a microwave digestion system, whilst mercury was analyzed by cold vapour atomic absorption spectrometry (CV–AAS). The content of mercury ranged from 0.018 mg/kg to 0.26 mg/kg in 2002, from 0.010 mg/kg to 0.42 mg/kg in 2005, from 0.010 mg/kg to 0.60 mg/kg in 2010 and from 0.020 mg/kg to 0.25 mg/kg in 2015. Analysis of the median values shows the increase of the content in the period 2002–2010 and a slight reduction of the air pollution with Hg in the period 2010–2015. Mercury distribution maps show that sites with increased concentrations of mercury in moss are likely impacted by anthropogenic pollution. The results were compared to similar studies done during the same years in neighboring countries and in Norway—which is a pristine area and serves as a reference, and it was concluded that mercury air pollution in Macedonia is significant primarily in industrialized regions.

Keywords: air pollution; moss biomonitoring; potentially toxic elements; mercury; CV–AAS; Macedonia

1. Introduction

Air pollutants are considered all chemical compounds or elements released into the atmosphere that pose health hazards to ecosystems and humans [1]. The majority of potentially toxic elements (PTEs) originate mainly from anthropogenic sources [2–4]. Natural sources of these elements include volcanoes, forest fires, biological decomposition processes and oceans [4]. The largest anthropogenic sources of PTEs in the atmosphere are the combustion of fossil and biofuels, traffic and emissions from industrial processes [5,6].

Due to the specific features and the effects on human health, the discharge of mercury in the environment has been identified as a global problem [7]. Even the mercury is naturally occurring in the Earth's crust, the atmospheric emission is mostly in an elemental mercury vapour form. Mercury enters the environment through volcanic eruptions and erosion of natural mercury–containing deposits, but also from forest fires and uncontrolled coal bed fires [7]. Anthropogenic sources of Hg are mostly connected with extraction refining and use of fossil fuels, metal production (Cu, Zn, Pb, etc.), production of inorganic materials (cement, paper production), recycling processes, chlorine–alkaline processes, etc. [7] High concentrations of mercury in human organism impact the central nervous system, especially the sensory, auditory and visual, parts of the brain that can affect coordination, lower immunity, heart attack risk, nervous system damage, and impair reproduction [8,9]. Therefore, timely and reliable identification of mercury discharge and presence in the environment is crucial.

Estimation of atmospheric heavy metal deposition using carpet–forming moss was done for the first time in the 1960s by Rühling and Tyler [10]. The moss analysis technique provides an alternative, time–integrated measure of the spatial patterns of PTEs deposition from the atmosphere to terrestrial ecosystems. The technique avoids the need for deploying large numbers of deposition collectors with an associated long–term programme of routine sample collection. Since 1990, the European moss survey has been repeated at five–yearly intervals [5,6,11,12] and the latest survey was conducted in 2015 with more than 35 participating countries [13]. The European moss survey gives data on the contents of ten potentially toxic metals (As, Cd, Cr, Cu, Fe, Hg, Ni, Pb, V, Zn) as well as the content of nitrogen [13–16].

The first study of mercury air pollution on the whole territory of Macedonia, using moss species as biomonitors, was undertaken in 2002, and the study was repeated in 2005, 2010, and 2015 within the European moss survey [17–23]. Pollution of soil samples with Hg, obtained in the vicinity of town Veles, due to work of the lead and zinc metallurgical plant situated near the town, was studied by Stafilov et al. [24,25] showing the increase of the content of Hg in topsoil over the European Hg average by a factor of 3.2. Analysis of the 174 soil samples collected in Kavadarci, an area in the south of the country [26,27] showed both lithogenic and anthropogenic influence of mercury in the collected samples. In some area along the river Vardar and the city of Kavadarci, the content of Hg was up to 3.8 mg/kg. Increased content of Hg was also found in soil samples in the vicinity of the "Allchar" As–Sb–Tl mine, located to the south of the city of Kavadarci [28]. This mine is lithogenic in origin but contributes to localized air pollution and it has been shown to influence the communities in this geographic area presenting natural phenomena.

The aims of this work are to investigate and present the temporal trends of the content of Hg in moss samples from the results of the surveys performed in the 2002–2015 period, to determine the places most affected by Hg pollution, and to try to connect the pollution with known anthropogenic activities in the regions, to distinguish natural from anthropogenic sources, to identify the deposition patterns and to compare results with previous studies in the neighboring countries and pristine areas.

2. Materials and Methods

2.1. Study Area

The Republic of Macedonia is a landlocked country situated in the central part of the Balkan Peninsula. Macedonia is bordering Serbia and Kosovo to the north, Bulgaria to the east, Albania to the west, and Greece to the south (Figure 1). A detailed description of the country (location, climate, and demographics) can be found elsewhere [17–23,29,30]. The location of main industrial activities and their input of different PTEs, has been also previously reported in several studies [31–46].

2.2. Sampling Sample Preparation and Instrumentation

In 2002, 2005, 2010 and 2015 moss samples from two moss species were collected on the entire territory of Macedonia [17–23]. In total, 72 moss samples were collected in each of the sampling campaigns (Figure 2). Collected moss samples were from the two most abundant moss species (*Hypnum cupressiforme* and *Homalothecium lutescens*). In locations where the two–moss species were collected, the interspecies comparison showed no differences within error estimates. The sampling procedure followed the principles of European moss surveys [12,47–49]. The moss samples were previously digested by microwave digestion system at 180 °C with nitric acid and then mercury was determined by cold vapor atomic absorption spectrometry (CV–AAS) [50].

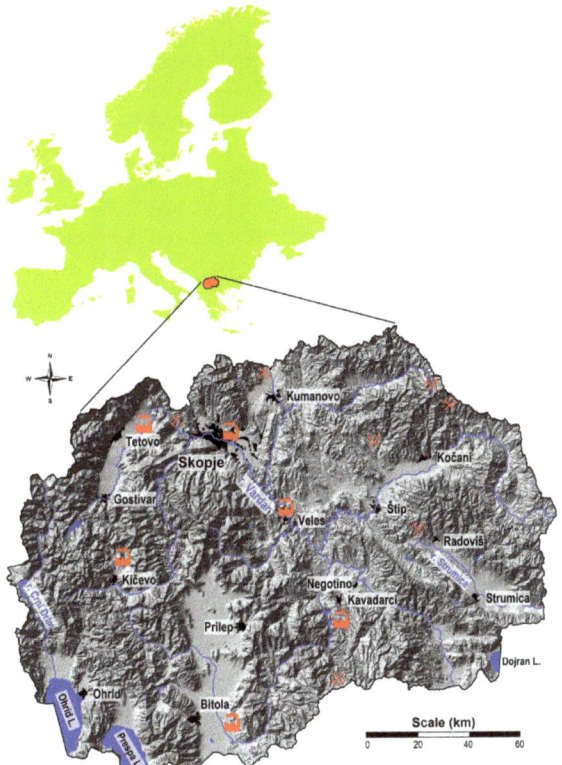

Figure 1. Map of the Republic of Macedonia.

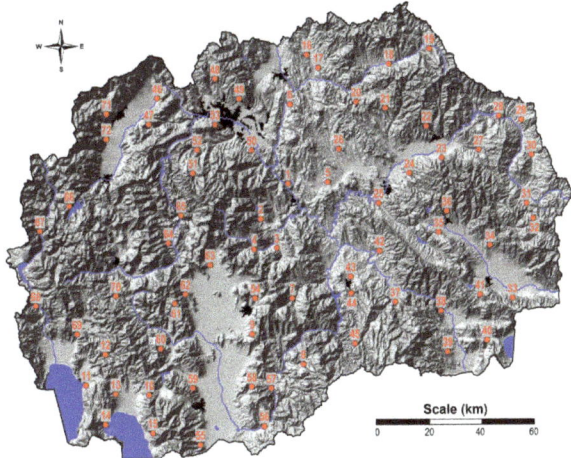

Figure 2. Location of sampling points.

2.3. *Quality Control*

The quality control of Hg determination was ensured by standard reference materials M2 and M3, which are prepared for the European Moss Survey [50,51] as well as the standard addition method.

2.4. Statistical Methods

Statistical analysis of the obtained data from the studies were made using the statistical software Statistica 13 (StatSoft, Inc., Tulsa, OK, USA) [52,53]. The common universal kriging with linear variogram interpolation method was applied to construct the maps of mercury areal distribution [54]. The basic grid cell size for interpolation was 1 × 1 km. For class limits the percentile values of distribution of all interpolated values (2002/2005/2010/2015) were chosen. Seven classes of the following percentile values were selected: 0–10, 10–25, 25–40, 40–60, 60–75, 75–90 and 90–100. In addition, an analysis of variance (ANOVA) was performed which showed significant differences between the four sampling seasons.

3. Results and Discussion

Results of the descriptive statistics of mercury content in the moss samples collected in 2002, 2005, 2010, and 2015 are given in Table 1. In Table 2, the obtained results were compared with the results of the content off Hg in moss samples obtained from similar studies in the neighbouring countries, as well as Norway which is considered a pristine area. The maps of the spatial distribution of Hg in moss samples collected in 2002, 2005, 2010, and 2015 to observe the trends of pollution in Macedonia during this period are given in Figures 3 and 4.

Table 1. Descriptive statistics of measurement according to sampling campaign, (in mg/kg).

	N	X_a	Md	Min	Max	P_{10}	P_{90}	S	CV	A	E
2002	72	0.069	0.056	0.018	0.26	0.034	0.114	0.040	60	2.03	7.08
2005	72	0.080	0.068	0.010	0.42	0.012	0.14	0.072	89	2.40	7.69
2010	72	0.110	0.093	0.010	0.60	0.05	0.16	0.094	84	3.57	14.2
2015	72	0.087	0.084	0.020	0.25	0.020	0.15	0.050	58	0.67	0.75

N—number of samples, X_a—arithmetic mean, Md—median, Min—minimum, Max—maximum, P_{10}—10 percentile, P_{90}—90 percentile, S—standard deviation, CV—coefficient of variation, A—asymmetry, E—distribution.

Table 2. The median values and ranges for the content of Hg obtained for Macedonia, neighbouring countries, and Norway.

	No. of Samples	Median (mg/kg)	Range (mg/kg)
Macedonia, 2002	72	0.056	0.018–0.26
Serbia, 2000 [55]	92	0.386	0.01–2.69
Norway, 2000 [11]	464	0.052	0.022–0.208
Macedonia, 2005	72	0.068	0.010–0.42
Croatia, 2006 [50]	94	0.064	0.007–0.301
Slovenia, 2005 [48]	57	0.095	0.050–0.175
Norway, 2005 [48]	100	0.046	0.026–0.166
Macedonia, 2010	72	0.093	0.010–0.60
Croatia, 2010 [56]	121	0.043	0.010–0.145
Slovenia, 2010 [6]	63	0.056	0.030–0.16
Albania 2010 [6]	59	0.130	0.031–2.23
Kosovo 2010 [6]	25	0.033	0.009–0.35
Norway, 2010 [6]	463	0.060	<0.024–0.34
Macedonia, 2015	72	0.084	0.020–0.25
Albania, 2015 [57]	55	0.049	0.006–0.21
Norway, 2015 [58]	229	0.050	0.005–0.53

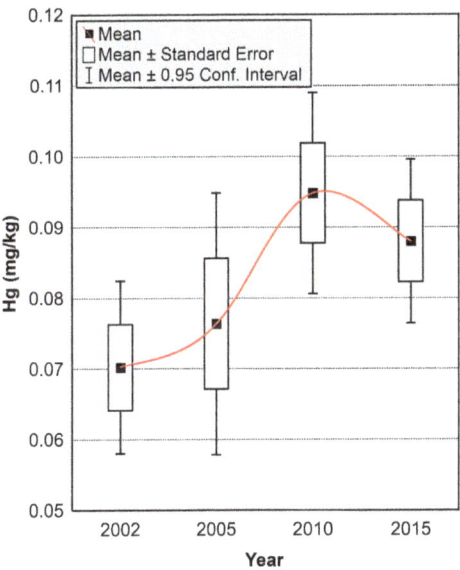

Figure 3. Box plots of Hg according to the year of sampling.

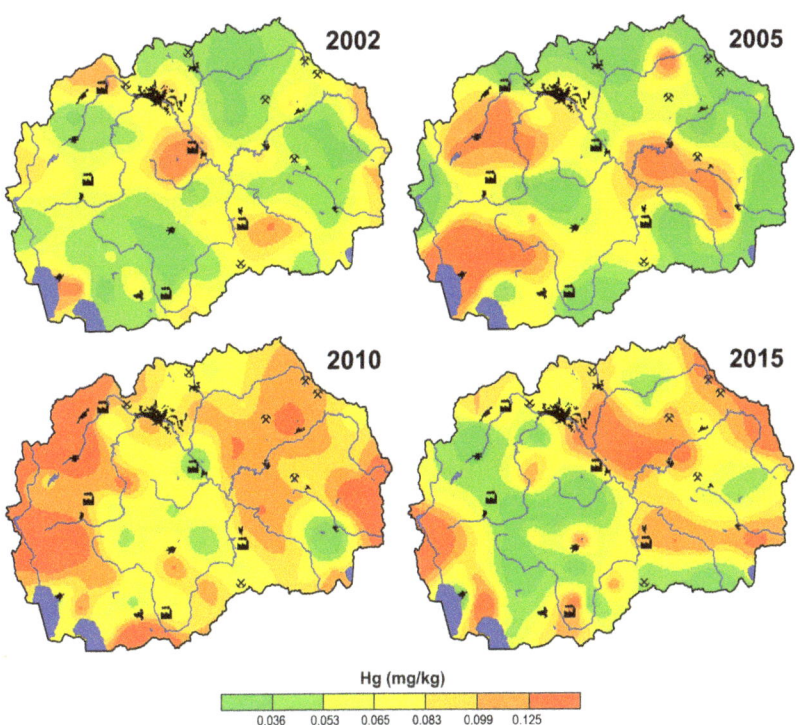

Figure 4. Distribution of Hg in moss samples collected in 2002, 2005, 2010 and 2015.

The mean value of mercury content in the moss samples increased in the period 2002–2010, while the value decreased for the samples collected in 2015 (Table 2, Figure 3). The coefficients of variation (CV%) are very high, with high values of asymmetry and distribution with a large range of variation of the positively skewed concentration data indicating the influence of different natural and anthropogenic factors. In addition, an analysis of variance (ANOVA) was performed which showed significant differences between the four sampling seasons (F = 5.43/p = 0.0012), which means that moss biomonitoring every 5 years is appropriate.

From the results of median values and the ranges of the content of Hg in Macedonia and those obtained in some neighbouring countries and Norway (Table 2), it can be seen that the median value obtained in 2002 (0.056 mg/kg) is slightly higher than the median value obtained from the survey in the same year in Norway [58] (which is usually considered as a pristine area [11]) (0.052 mg/kg) and 9.9 times lower than those obtained for Serbia [55]. The median value obtained for the 2005 study is comparable with the median value obtained in the similar study done in Croatia [56], lower than the median value for Slovenia [48] and 1, 5 times higher in comparison with the median value for Norway [48]. The obtained median value for 2010 is higher than the values obtained in the neighbouring countries except in the case of Albania [15], which is the case for data for 2015 as well. It should be also emphasized that the maximal value for the content of Hg in moss samples from Macedonia in 2002, 2005, and 2010 is higher than the maximal content of Hg in moss samples from Norway, while in 2015 is twice smaller.

Distribution maps based on the content of Hg in moss samples collected in 2002, 2005, 2010 and 2015 (Figure 4) show some differences and a discrepancy, i.e., variation with time as well as geographically, but it can be noticed that an increased Hg content is visible in the same regions for the whole period of the research. In 2002, the higher Hg content in the moss samples can be seen along the valley of the river Vardar, which extends from the cities of Skopje to Veles and continues southeast toward the border with Greece. It should be also mentioned that in 2002 pollution with Hg is connected with the activities of the lead and zinc smelter plant in the town of Veles [25] which had operated until 2003, as well with the work of the steelwork in Skopje [17,45]. Higher content of Hg in the mosses collected in 2002 sampling campaign was observed on the north–eastern part of the country (in the vicinity of the town of Kočani) where three Pb–Zn mines and flotation plants (Sasa, Toranica, Zletovo) are located [39].

In 2005 the enrichment of Hg was connected with the operation of the steelwork in Skopje [18] as well as the pollution by the lead and zinc smelter plant in Veles despite its closure in 2003. In contrast with the moss survey in 2002, enrichment with Hg in 2005 is also observed in the areas of Bitola in the south–western part of the country and in the west–central part of the country near the town of Kičevo, where two thermoelectric power plants using not pre–treated i.e. pre–washed prior combustion coal–lignite for energy production are located. Higher content of Hg in the mosses is also noticed in the eastern part of the country, near the town of Radoviš where copper mine and flotation were reactivated in this period.

With the reactivation of Pb–Zn mines (Toranica, Zletovo, and Sasa), in 2010 the enrichment of the content of Hg was observed in the eastern part of Macedonia. In the period of 2001/02 until 2006/07 these mines were not active, but in 2006 (Sasa) and 2007 (Toranica, Zletovo) were reactivated. In the mining area, millions of tons of flotation tailings are deposited on the ground, and can easily be dispersed into the biosphere by wind [39–41]. Furthermore, the enrichment in 2010 is also observed in the area of the steelwork in Skopje, near Bitola, Kičevo, but also near Kavadarci which is related to reactivation and the increasing of the production capacity of the ferro–nickel smelter in 2004 [19]. Higher content of Hg in the mosses collected near the western border of the country can be explained by the transboundary transport from Albania where the content of Hg in the mosses in this period was substantially higher than the samples collected in Macedonia [57].

Although the median value of Hg content in the mosses in Macedonia in 2015 is lower than the one in 2010, the high values are observed in the same areas as in the previous years.

High enrichment of the moss samples collected near the capital of Skopje, is partially connected with the presence of mercury–contaminated sites at the former "Organic Chemical Industry in Skopje

–OHIS" chlor–alkali plant situated in the town (operational from 1964 until 1995). The plant was based on mercury cell electrolysis and during the whole operational period, sodium hydroxide and hydrochloric acid were produced. It is assumed that part of the waste arising from the chlorine production containing mercury was dumped onto the mixed waste dump within the plant surrounding and that 400 tons of mercury were lost in the environment [59].

The results from this study and from previous work [24–27], in which the content of Hg in various environmental samples (soil, water, sediments) were determined, has been used to prepare the Minamata Initial Assessment in the Republic of Macedonia and the National inventory on mercury releases developed in 2017 [59]. The reference base year is 2013 and data for this year have been used in the inventory when available. The National mercury inventory identified most of the sources of mercury releases in the country and estimates or quantifies the releases which are consistent with the results obtained from these studies using moss biomonitoring.

4. Conclusions

The content of mercury was determined in moss samples collected in 2002, 2005, 2010, and 2015 all over the territory of Macedonia by using CV–AAS. ANOVA analysis of the results showed significant difference between the results obtained in the four sampling seasons. Although distribution maps show some differences and a discrepancy, the largest anthropogenic impact of air pollution with mercury was found near the abandoned lead–zinc smelter in the town of Veles, lead and zinc mines Sasa, Zletovo and Toranica – in the north–eastern part of the country, ferronickel smelter at the vicinity of the town of Kavadarci, and the copper mine and flotation near the town of Radoviš. The high content of mercury in the moss samples was also observed near the two thermoelectric power plants located in the vicinity of the towns of Bitola and Kičevo. Evidence of transboundary transport from Albania was observed in the western part of the country in samples collected in 2010 and 2015. Elevated values of Hg content in the moss samples collected near Skopje can be explained with the work of the former chlor–alkali plant "OHIS" and the inappropriate deposition of mercury–contaminated waste in the vicinity of this plant. This work is essential for modeling the mercury pollution in Macedonia, as well as monitoring the future trends aiming to preserve the quality of the ecosystems from deteriorating.

Author Contributions: Conceptualization, L.B. and T.S.; methodology, T.S.; software, R.Š.; investigation, T.S.; L.B. and K.B.A.; data curation, R.Š.; writing—original draft preparation, L.B.; writing—review and editing, L.B. and T.S.; visualization, R.Š. All authors have read and agreed to the published version of the manuscript.

Funding: This research received no external funding.

Conflicts of Interest: The authors declare no conflict of interest.

References

1. Moriarty, F. *Ecotoxicology—The Study of Pollutants in Ecosystems*, 3rd ed.; Academic Press: San Diego, CA, USA, 1999.
2. Whelpdale, D.M.; Summers, P.W.; Sanhueza, E. A global overview of atmospheric acid deposition fluxes. *Environ. Monit. Assess.* **1998**, *48*, 217–247. [CrossRef]
3. Pacyna, J.M.; Pacyna, E.G. An assessment of global and regional emissions of trace metals to the atmosphere from anthropogenic sources worldwide. *Environ. Rev.* **2001**, *9*, 269–298. [CrossRef]
4. Nriagu, J.O. A global assessment of natural sources of atmospheric trace metals. *Nature* **1999**, *338*, 47–49. [CrossRef]
5. Harmens, H.; Norris, D.A.; Steinnes, E.; Kubin, E.; Piispanen, J.; Alber, R.; Aleksiayenak, Y.; Blum, O.; Coşkun, M.; Dam, M.; et al. Mosses as biomonitors of atmospheric heavy metal deposition: Spatial and temporal trends in Europe. *Environ. Pollut.* **2010**, *158*, 3144–3156. [CrossRef]
6. Harmens, H.; Norris, D.A.; Sharps, K.; Mills, G.; Alber, R.; Aleksiayenak, Y.; Blum, O.; Cucu-Man, S.M.; Dam, M.; De Temmerman, L.; et al. Heavy metal and nitrogen concentrations in mosses are declining across Europe whilst some "hotspots" remain in 2010. *Environ. Pollut.* **2015**, *200*, 93–104. [CrossRef]

7. Pirrone, N.; Cinnirella, S.; Feng, X.; Finkelman, R.B.; Friedli, H.R.; Leaner, J.; Mason, R.; Mukherjee, A.B.; Stracher, G.B.; Streets, D.G.; et al. Global mercury emissions to the atmosphere from anthropogenic and natural sources. *Atmos. Chem. Phys.* **2010**, *10*, 5951–5964. [CrossRef]
8. Clarkson, T.W. The toxicology of mercury. *Crit. Rev. Toxicol.* **1997**, *34*, 369–403. [CrossRef]
9. Tchounwou, P.B.; Ayensu, W.K.; Ninashvili, N.; Sutto, D. Environmental exposure to mercury and its toxicopathologic implications for public health. *Environ. Toxic.* **2003**, *18*, 149–175. [CrossRef]
10. Rühling, Å.; Tyler, G. An ecological approach to the lead problem. *Bot. Not.* **1968**, *122*, 248–342.
11. Harmens, H.; Buse, A.; Büker, P.; Norris, D.; Mills, G.; Williams, B. Heavy metal concentrations in European mosses: 2000/2001 survey. *J. Atmos. Chem.* **2004**, *49*, 425–436. [CrossRef]
12. Rühling, Å.; Steinnes, E. *Atmospheric Heavy Metal Deposition in Europe 1995–1996*. NORD 1998:15; Nordic Council of Ministers: Copenhagen, Denmark, 1998.
13. Frontasyeva, M.; Harmens, H.; Uzhinskiy, A.; Chaligava, O.; Aničić Urošević, M.; Vergel, K.; Tepanosyan, G.; Steinnes, E.; Aleksiayenak, Y.; Tabors, G.; et al. *Mosses as Biomonitors of Air Pollution: 2015/2016 Survey on Heavy Metals, Nitrogen and POPs in Europe and Beyond*; Joint Institute for Nuclear Research JINR: Dubna, Russia, 2020.
14. Harmens, H.; Norris, D.A.; Koerber, G.R.; Buse, A.; Steinnes, E.; Rühling, Å. Temporal trends in the concentration of arsenic, chromium, copper, iron, nickel, vanadium and zinc in mosses across Europe between 1990 and 2000. *Atmos. Environ.* **2007**, *41*, 6673–6687. [CrossRef]
15. Harmens, H.; Norris, D.A.; Koerber, G.R.; Buse, A.; Steinnes, E.; Rühling, Å. Temporal trends (1990x2000) in the concentration of cadmium, lead and mercury in mosses across Europe. *Environ. Poll.* **2008**, *151*, 368–376. [CrossRef]
16. Nickel, S.; Schröder, W.; Schmalfuss, R.; Saathoff, M.; Harmens, H.; Mills, G.; Frontasyeva, M.V.; Barandovski, L.; Blum, O.; Carballeira, A.; et al. Modelling spatial patterns of correlations between concentrations of heavy metals in mosses and atmospheric deposition in 2010 across Europe. *Environ. Sci. Eur.* **2018**, *30*, 53. [CrossRef]
17. Barandovski, L.; Cekova, M.; Frontasyeva, M.V.; Pavlov, S.S.; Stafilov, T.; Steinnes, E.; Urumov, V. Atmospheric deposition of trace element pollutants in Macedonia studied by the moss biomonitoring technique. *Environ. Monit. Assess.* **2008**, *138*, 107–118. [CrossRef]
18. Barandovski, L.; Frontasyeva, M.V.; Stafilov, T.; Šajn, R.; Pavlov, S.; Enimiteva, V. Trends of atmospheric deposition of trace elements in Macedonia by the moss biomonitoring technique. *J. Environ. Sci. Health A* **2012**, *47*, 2000–2015. [CrossRef]
19. Barandovski, L.; Stafilov, T.; Šajn, R.; Frontasyeva, M.V.; Bačeva, K. Air pollution study in Macedonia using a moss biomonitoring technique, ICP–AES, and AAS. *Maced. J. Chem. Chem. Eng.* **2013**, *32*, 89–107. [CrossRef]
20. Barandovski, L.; Frontasyeva, M.V.; Stafilov, T.; Šajn, R.; Ostrovnaya, T.M. Multi–element atmospheric deposition in Macedonia studied by the moss biomonitoring technique. *Environ. Sci. Pollut. Res.* **2015**, *22*, 16077–16097. [CrossRef] [PubMed]
21. Stafilov, T.; Šajn, R.; Barandovski, L.; Bačeva, K.A.; Malinovska, S. Moss biomonitoring of atmospheric deposition study of minor and trace elements in Macedonia. *Air Qual. Atmos. Health* **2018**, *11*, 137–152. [CrossRef]
22. Stafilov, T.; Špirić, Z.; Glad, M.; Barandovski, L.; Andonovska, K.B.; Šajn, R.; Antonić, O. Study of nitrogen pollution in the Republic of North Macedonia by moss biomonitoring and Kjeldahl method. *J. Environ. Sci. Health A* **2020**, *55*, 759–764. [CrossRef] [PubMed]
23. Barandovski, L.; Stafilov, T.; Šajn, R.; Frontasyeva, M.; Bačeva Andonovska, K. Atmospheric heavy metal deposition in North Macedonia from 2002 to 2010 studied by the moss biomonitoring technique. *Atmosphere* **2020**, *11*, 929. [CrossRef]
24. Stafilov, T.; Šajn, R.; Pančevski, Z.; Boev, B.; Frontasyeva, M.V.; Strelkova, L.P. *Geochemical Atlas of Veles and the Environs*; Faculty of Natural Sciences and Mathematics: Skopje, Macedonia, 2008.
25. Stafilov, T.; Šajn, R.; Pančevski, Z.; Boev, B.; Frontasyeva, M.V.; Strelkova, L.P. Heavy metal contamination of surface soils around a lead and zinc smelter in the Republic of Macedonia. *J. Hazard. Mater.* **2010**, *175*, 896–914. [CrossRef]
26. Stafilov, T.; Šajn, R.; Boev, B.; Cvetković, J.; Mukaetov, D.; Andreevski, M. *Geochemical Atlas of Kavadarci and the Environs*; Faculty of Natural Sciences and Mathematics: Skopje, Macedonia, 2008.

27. Stafilov, T.; Šajn, R.; Boev, B.; Cvetković, J.; Mukaetov, D.; Andreevski, M.; Lepitkova, S. Distribution of some elements in surface soil over the Kavadarci Region, Republic of Macedonia. *Environ. Earth Sci.* **2010**, *61*, 1515–1530. [CrossRef]
28. Bačeva, K.; Stafilov, T.; Šajn, R.; Tănăselia, C.; Makreski, P. Distribution of chemical elements in soils and stream sediments in the area of abandoned Sb–As–Tl Allchar mine, Republic of Macedonia. *Environ. Res.* **2014**, *133*, 77–89. [CrossRef] [PubMed]
29. State Statistical Office of the Republic of Macedonia. Environmental Statistics. 2019. Available online: http://www.stat.gov.mk/Publikacii/ZivotnaSredina2019.pdf (accessed on 15 November 2020).
30. Lazarevski, A. *Climate in Macedonia*; Kultura: Skopje, Macedonia, 1993. (In Macedonian)
31. Balabanova, B.; Stafilov, T.; Bačeva, K.; Šajn, R. Biomonitoring of atmospheric pollution with heavy metals in the copper mine vicinity located near Radoviš, Republic of Macedonia. *J. Environ. Sci. Health A* **2010**, *45*, 1504–1518. [CrossRef]
32. Balabanova, B.; Stafilov, T.; Šajn, R.; Bačeva, K. Distribution of chemical elements in attic dust as reflection of lithology and anthropogenic influence in the vicinity of copper mine and flotation. *Arch. Environ. Contam. Toxicol.* **2011**, *6*, 173–184. [CrossRef] [PubMed]
33. Bačeva, K.; Stafilov, T.; Šajn, R.; Tănăselia, C.; Ilić Popov, S. Distribution of chemical elements in attic dust in the vicinity of ferronickel smelter plant. *Fresenius Environ. Bull.* **2011**, *20*, 2306–2314.
34. Bačeva, K.; Stafilov, T.; Šajn, R.; Tănăselia, C. Moss biomonitoring of air pollution with heavy metals in the vicinity of a ferronickel smelter plant. *J. Environ. Sci. Health A* **2012**, *47*, 645–656. [CrossRef]
35. Bačeva, K.; Stafilov, T.; Šajn, R.; Tănăselia, C. Air dispersion of heavy metals in vicinity of the As–Sb–Tl abounded mine and responsiveness of moss as a biomonitoring media in small scale investigations. *Environ. Sci. Pollut. Res.* **2013**, *20*, 8763–8779. [CrossRef]
36. Stafilov, T. Environmental pollution with heavy metals in the Republic of Macedonia. *Contrib. Sect. Nat. Math. Biotechnol. Sci. MASA* **2014**, *35*, 81–119. [CrossRef]
37. Balabanova, B.; Stafilov, T.; Šajn, R.; Bačeva, K. Comparison of response of moss, lichens and attic dust to geology and atmospheric pollution from copper mine. *Int. J. Environ. Sci. Technol.* **2014**, *11*, 517–528. [CrossRef]
38. Angelovska, S.; Stafilov, T.; Šajn, R.; Bačeva, K.; Balabanova, B. Moss biomonitoring of air pollution with heavy metals in the vicinity of Pb–Zn mine "Toranica" near the town of Kriva Palanka. *Modern Chem. Appl.* **2014**, *2*, 1–6.
39. Balabanova, B.; Stafilov, T.; Šajn, R.; Tănăselia, C. Multivariate extraction of dominant geochemical markers for 69 elements deposition in Bregalnica River Basin, Republic of Macedonia (moss biomonitoring). *Environ. Sci. Poll. Res.* **2016**, *23*, 22852–22870. [CrossRef] [PubMed]
40. Balabanova, B.; Stafilov, T.; Šajn, R.; Bačeva Andonovska, K. Quantitative assessment of metal elements using moss species as biomonitors in downwind area of lead-zinc mine. *J. Environ. Sci. Health A* **2017**, *52*, 290–301.
41. Stafilov, T.; Šajn, R. *Geochemical Atlas of the Republic of Macedonia*; Faculty of Natural Sciences and Mathematics, Sc. Cyril and Methodius University: Skopje, Macedonia, 2016.
42. Stafilov, T.; Šajn, R.; Arapčeska, M.; Kungulovski, I.; Alijagić, J. Geochemical properties of topsoil around the coal mine and thermoelectric power plant. *J. Environ. Sci. Health A* **2018**, *53*, 793–808.
43. Stafilov, T.; Šajn, R.; Sulejmani, F.; Bačeva, K. Geochemical properties of topsoil around the open coal mine and Oslomej thermoelectric power plant R.; Macedonia. *Geol. Croat.* **2014**, *67*, 33–44. [CrossRef]
44. Balabanova, B.; Stafilov, T.; Šajn, R. Enchasing anthropogenic element trackers for evidence of long-term atmospheric depositions in mine environ. *J. Environ. Sci. Health A* **2019**, *54*, 988–998.
45. Stafilov, T.; Šajn, R.; Ahmeti, L. Geochemical characteristics of soil of the city of Skopje, Republic of Macedonia. *J. Environ. Sci. Health A* **2019**, *54*, 972–987.
46. Stafilov, T.; Šajn, R. Spatial distribution and pollution assessment of heavy metals in soil from the Republic of North Macedonia. *J. Environ. Sci. Health A* **2019**, *54*, 1457–1474.
47. Harmens, H.; Mills, G.; Hayes, F.; Williams, P.; Soja, G.; Riss, A.; Zechmeister, H.G.; Temmerman, L.D.; Yurukova, L.; Stamenov, J.; et al. *Air pollution and Vegetation, ICP Vegetation Annual Report 2003/2004*; ICP Vegetation Coordination Centre: Bangor, UK, 2004.
48. Harmens, H.; Mills, G.; Hayes, F.; Williams, P.; Temmerman, L.D.; Soja, G.; Stabentheiner, E.; Riss, A.; Zechmeister, H.G.; Zhuk, I.; et al. *Air Pollution and Vegetation ICP Vegetation Annual Report 2004/2005*; ICP Vegetation Coordination Centre: Bangor, UK, 2005.

49. Harmens, H.; Mills, G.; Hayes, F.; Norris, D.; Lazo, P.; Soja, G.; Zechmeister, H.; Aleksiayenek, Y.; Temmerman, L.D.; Vandermeiren, K.; et al. *Air Pollution and Vegetation ICP Vegetation Annual Report 2011/2012*; ICP Vegetation Coordination Centre: Bangor, UK, 2012.
50. Špirić, Z.; Vučković, I.; Stafilov, T.; Kušan, V.; Bačeva, K. Biomonitoring of air pollution with mercury in Croatia by using moss species and CV–AAS. *Environ. Monit. Assess.* **2014**, *186*, 4357–4366. [CrossRef]
51. Steinnes, E.; Rühling, Å.; Lippo, H.; Mäkinen, A. Reference material for large-scale metal deposition surveys. *Accredit. Qual. Assur.* **1997**, *2*, 243–249.
52. Snedecor, G.W.; Cochran, W.G. *Statistical Methods*; The Iowa State University Press: Ames, IA, USA, 1967.
53. Hollander, M.; Wolfe, D.A. *Nonparametric Statistical Methods*, 2nd ed.; Wiley: New York, NY, USA, 1999.
54. Davis, J.C. *Statistic and Data Analysis in Geology*; Willey: New York, NY, USA, 1986; p. 646.
55. Frontasyeva, M.V.; Galinskaya, T.Y.; Krmar, M.; Matavulj, M.; Pavlov, S.S.; Radnović, D.; Steinnes, E. Atmospheric deposition of heavy metals in northern Serbia and Bosnia-Herzegovina studied by moss biomonitoring, neutron activation analysis and GIS technology. *J. Radioanal. Nucl. Chem.* **2004**, *259*, 141–147. [CrossRef]
56. Špirić, Z.; Frontasyeva, M.; Steinnes, E.; Stafilov, T. Multi–element atmospheric deposition study in Croatia. *Int. J. Environ. Anal. Chem.* **2012**, *92*, 1200–1214. [CrossRef]
57. Lazo, P.; Stafilov, T.; Qarri, F.; Allajbeu, S.; Bekteshi, L.; Frontasyeva, M.; Harmens, H. Spatial distribution and temporal trend of airborne trace metal deposition in Albania studied by moss biomonitoring. *Ecol. Indic.* **2019**, *101*, 1007–1017. [CrossRef]
58. Steinnes, E.; Uggerud, H.T.; Pfaffhuber, K.A.; Berg, T. *Atmospheric Deposition of Heavy Metals in Norway—National Moss Survey 2015. M–595*; Norwegian Institute for Air Research: Trondheim, Norway, 2015.
59. Stafilov, T.; Mickovski, A.; Mihajlov, M. *National Inventory of Mercury Releases in the Republic of Macedonia*; Ministry of Environment and Physical Planning: Skopje, Macedonia, 2017.

Publisher's Note: MDPI stays neutral with regard to jurisdictional claims in published maps and institutional affiliations.

© 2020 by the authors. Licensee MDPI, Basel, Switzerland. This article is an open access article distributed under the terms and conditions of the Creative Commons Attribution (CC BY) license (http://creativecommons.org/licenses/by/4.0/).

Article

Impacts of Anthropogenic Emissions and Meteorological Variation on Hg Wet Deposition in Chongming, China

Yi Tang [1], Qingru Wu [1,2,*], Wei Gao [3], Shuxiao Wang [1,2], Zhijian Li [1], Kaiyun Liu [1] and Deming Han [1]

[1] State Key Joint Laboratory of Environmental Simulation and Pollution Control, School of Environment, Tsinghua University, Beijing 100084, China; tangy19@mails.tsinghua.edu.cn (Y.T.); shxwang@tsinghua.edu.cn (S.W.); yit236717@gmail.com (Z.L.); liuky16@mails.tsinghua.edu.cn (K.L.); handeem@tsinghua.edu.cn (D.H.)
[2] State Environmental Protection Key Laboratory of Sources and Control of Air Pollution Complex, Beijing 100084, China
[3] Yangtze River Delta Center for Environmental Meteorology Prediction and Warning, Shanghai 20030, China; gao9989@gmail.com
* Correspondence: qrwu@tsinghua.edu.cn

Received: 15 October 2020; Accepted: 27 November 2020; Published: 30 November 2020

Abstract: Mercury (Hg) is a ubiquitous environmental toxicant that has caused global concern due to its persistence and bioaccumulation in the environment. Wet deposition is a crucial Hg input for both terrestrial and aquatic environments and is a significant indicator for evaluating the effectiveness of anthropogenic Hg control. Rainwater samples were collected from May 2014 to October 2018 in Chongming Island to understand the multi-year Hg wet deposition characteristics. The annual Hg wet deposition flux ranged from 2.6 to 9.8 $\mu g\ m^{-2}\ yr^{-1}$ (mean: 4.9 $\mu g\ m^{-2}\ yr^{-1}$). Hg wet deposition flux in Chongming was comparable to the observations at temperate and subtropical background sites (2.0–10.2 $\mu g\ m^{-2}\ yr^{-1}$) in the northern hemisphere. Hg wet deposition flux decreased from 8.6 $\mu g\ m^{-2}\ yr^{-1}$ in 2014–2015 to 3.8 $\mu g\ m^{-2}\ yr^{-1}$ in 2016 and was attributed to a decrease in the volume-weighted mean (VWM) Hg concentration (−4.1 $ng\ L^{-1}\ yr^{-1}$). The reduced VWM Hg was explained by the decreasing atmospheric Hg and anthropogenic emissions reductions. The annual Hg wet deposition flux further decreased from 3.8 $\mu g\ m^{-2}$ in 2016 to 2.6 $\mu g\ m^{-2}$ in 2018. The reduction of warm season (April–September) rainfall amounts (356–845 mm) mainly contributed to the Hg wet deposition flux reduction during 2016–2018. The multi-year monitoring results suggest that long-term measurements are necessary when using wet deposition as an indicator to reflect the impact of anthropogenic efforts on mercury pollution control and meteorological condition variations.

Keywords: Hg wet deposition flux; VWM Hg concentrations; Chongming; anthropogenic emissions; meteorological condition

---

1. Introduction

Mercury (Hg) is a pollutant of global concern due to its long residence time and neurotoxicity. Once emitted in the atmosphere, Hg can be transported in long distances and cause ecological damage globally [1]. Atmospheric Hg exists in three forms: gaseous elemental mercury (GEM), gaseous oxidized mercury (GOM), and particulate-bound mercury (PBM). GEM contributes to 95–99% of atmospheric Hg with a residence time of 0.5–2 years [2]. Although GOM and PBM represent less than 5% of atmospheric Hg, they impact the global Hg cycling by rapid dry deposition and wet scavenging such as rainfall [3,4]. Hg deposition could represent its pollution characteristics and help us understand Hg cycling. To date, Hg dry deposition is often estimated using measured atmospheric speciated mercury and meteorological parameters due to a lack of direct and accurate measurement methodologies [5,6]. The uncertainties of estimated Hg dry deposition can reach 50–200% or more, while the measurement of Hg wet deposition was thought to be relatively accurate with only experimental bias [6,7]. Thus, the investigation of Hg deposition on terrestrial and aquatic surfaces is often based on wet deposition measurements [8].

Hg wet deposition measurement has been conducted by the National Atmospheric Deposition Program (NADP) and Global Mercury Observation System (GMOS) in Europe and North America since the 1990s [9]. In 2015, GMOS and NADP have conducted ground-based monitoring at 35 and >50 stations around the world, respectively [10]. Previous studies have suggested that meteorological condition and anthropogenic emissions are the most important factors affecting Hg wet deposition flux [8,9]. The spatial difference of 17 GMOS sites in Hg wet deposition flux highlights the importance of rainfall amounts [9]. Long-term Hg wet deposition flux has demonstrated a significant decreasing trend since the 1990s at most European and North American observation sites [11,12]. The observed atmospheric Hg deposition downward trends have been explained by reduced anthropogenic Hg emissions and commercial product releases [12,13]. The similar trend of anthropogenic emission, atmospheric Hg concentrations, and Hg wet deposition illustrates the importance of anthropogenic emissions on Hg wet deposition [12].

Hg wet deposition flux in China has been observed since 2000 at various remote and urban sites, such as Bayinbuluk, Mt. Changbai, Shangri'La, and Nanjing [8,14]. However, most previous observations in China generally aimed to estimate the wet deposition flux in a relatively short period (e.g., one year) [8]. Recently, with the entry into force of the *Minamata Convention on Mercury*, it is possible that Hg wet deposition flux will be one indicator to evaluate the anthropogenic emission control effect of atmospheric mercury. Anthropogenic Hg emissions in China have also shown a decreasing trend during the past several years [15–17]. Therefore, we are interested to know how wet deposition flux of Hg responds to emission variations, which should be based on multi-year measurements.

In this study, we collected four years of rainwater samples on Chongming Island of Shanghai and calculated the Hg wet deposition flux. The latitudinal and seasonal patterns of Hg wet deposition flux distribution were studied to understand their spatiotemporal variation characteristics. The influences of anthropogenic emissions were investigated by analyzing backward trajectories and volume-weighted mean (VWM) Hg concentrations during different periods. In addition, rainfall and large-scale circulation were investigated to explain the trend of Hg wet deposition flux during 2016–2018 and reflect the impact of meteorological conditions.

2. Methodology

2.1. Site Description

The observation site is located at the top of the weather station at the Dongtan Birds National Natural Reserve, Chongming Island (31.52° N, 121.96° E, 10 m above sea level) (Figure S1). Chongming Island is located at the easternmost of the Yangtze River Delta (YRD) region, which comprised 240 million people and was responsible for 24% of China's GDP in 2018 [18]. Chongming Island has a subtropical monsoon climate with hot, humid summers and cold, dry winters. The dominant land-surface types in the surrounding 20 km are farmlands and wetlands. The downtown area of Shanghai is 50 km southwest of the sampling site. Thus, the wet deposition flux here could reflect the background concentrations of the YRD region and many atmospheric observations have been conducted at this station over the years due to its remote environment [19].

2.2. Rainwater Sampling and Hg Analysis

The collection of rainwater samples at Chongming in this study started in June 2014 and ended in October 2018. We encountered operational problems between June 2015 and February 2016, with no rainwater collected. The rainwater samples were collected every day from 08:00 to 08:00 LST (Local

Sidereal Time) using an automated rain collector (Hengda Company, type ZJV-3). The rainwater was collected in a PTFE (Polytetrafluoroethylene) bottle with a Teflon tube and placed in a refrigerator. The bottles were acid-cleaned and rinsed with deionized water before use. A 10 mL 3% HCl solution was placed into a bottle before sampling to store the Hg (II). Normally, we replaced 8 rainwater bottles every two weeks and recorded the data. The sampled bottles were transferred into the refrigerator until analysis. We prepared a whole experimental process blank to reduce experimental error by pouring deionized water into the rain collector and transferring it to the laboratory. The method recovery rate of 98.2 ± 8.2% was calculated by injecting 100 mL 100 ng L^{-1} $HgNO_3$ into the PTFE bottle and analyzing after two weeks.

The total Hg in the rainwater was analyzed by dual-gold-trap amalgamation and cold vapor atomic fluorescence spectrometry after BrCl oxidation, hydroxylamine hydrochloride neutralization, and $SnCl_2$ reduction. The analysis instrument was a Tekran 2600 using the EPA 1631 method. The detection limit of the Tekran 2600 was 0.1 ng L^{-1}. To confirm the consistency of the instrument, the $HgNO_3$ standards were used to prepare a five-point calibration (0, 5, 20, 50, 100 ng L^{-1}). The calibration curve is shown in Figure S2, with a regression coefficient of 0.997. All of the samples obtained results after subtracting the whole experimental process blank (0.2±0.4 ng L^{-1}).

2.3. Hg Wet Deposition Flux

The Hg wet deposition flux was calculated based on the mean VWM concentration and the total rainwater amount collected during the sampling period. The VWM concentration was calculated using the following equation:

$$VWM = \sum (C_i \times V_i) / \sum V_i, \tag{1}$$

where C_i is the Hg concentration for a daily rainfall event (ng L^{-1}), and V_i is the rainfall depth (mm). The wet deposition of Hg during a certain period was calculated by multiplying the sum of the rainfall depth during the calculation period by VWM concentration according to Equation (2):

$$F_w = VWM \times \sum_{i=1}^{i=n} V_i, \tag{2}$$

where F_w is the Hg wet deposition flux during a certain period. The monthly and annual Hg wet deposition flux was calculated based on monthly and annual rainfall depth and VWM concentrations to avoid the impact of missing values in daily rain events.

2.4. Analysis Method and Data Acquisition

To identify the influence of airmasses from different transport pathways, 48 h backward trajectories were estimated at 00:00 and 12:00 LST each rainy day using the Hybrid Single Particle Lagrangian Integrated Trajectory (HYSPLIT) model [20]. The 1° × 1° gridded meteorological data from the Global Data Assimilation System were used in this study. The endpoint was set at the sampling site with a height of 500 m above sea level, representing the center of the boundary layer [19].

Cluster analysis was used in the backward trajectories. In the procedure, each trajectory was coupled with the sampling concentrations (rainwater Hg and rainfall amounts) on that day. All of the calculation processes were carried out in the Meteoinfo software and ArcMap 10.2. More details about cluster analysis can be found in Dorling, et al. [21].

Atmospheric Hg data were obtained from the Tekran 2537X/1130/1135 speciation unit at the sampling site. The working conditions, quality assurance and quality control procedure of the instrument were described in Tang, et al. [22]. The total gaseous mercury (TGM) was the annual average during the sampling period, and Sen's slope was used to calculate the decreasing trend via the monthly VWM Hg data from the "mblm" package in R 4.0.2. To avoid the influence of missing values on interannual investigation, the Sen's slope of Apr–Sep and Oct–Mar were calculated and averaged for the total Sen's slope [23].

3. Results and Discussion

3.1. Characteristics of Hg Wet Deposition Flux

The daily VWM Hg concentrations ranged from 0.1 to 37.3 ng L^{-1} during the sampling period (Figure 1). The arithmetic mean value of VWM Hg concentrations (7.6 ng L^{-1}) was higher than the median value (5.7 ng L^{-1}), indicating the occurrence of episodic high concentration events. The highest frequencies were in the range from 2.0 to 6.0 ng L^{-1}. Our daily mean rainwater Hg data (7.5 ng L^{-1}) were within the range of values measured at other subtropical and temperate background monitoring sites in East Asia, such as Mt. Damei (3.7 ng L^{-1}), Mt. Ailao (3.7 ng L^{-1}), Mt. Changbai (7.4 ng L^{-1}), and Lulin (9.2 ng L^{-1}) (Table 1).

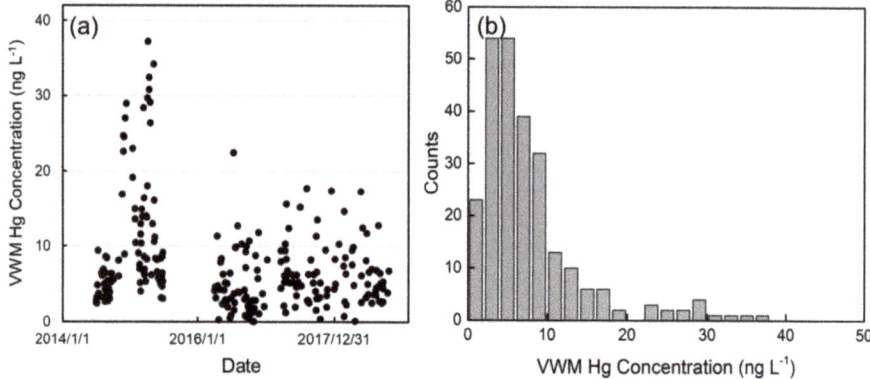

Figure 1. (a) Times series of rainy events and its volume-weighted mean (VWM) Hg concentrations. (b) The frequency distribution of VWM Hg concentrations.

Table 1. Summary of Hg wet deposition flux at various sites worldwide.

Name	Site Description	Longitude (°)	Latitude (°)	Period	Rain Depth (mm)	VWM Conc. (ng L^{-1})	Wet Deposition ($\mu g\ m^{-2}$)	Reference
Chongming	Remote	121.96	31.52	2014–2018	857	7.6	4.9	This study
Chongming	Remote	-	-	2008–2009	-	62	-	[24]
Lulin	Remote	120.87	23.47	2010–2013	3421	9.2	32.3	[25]
Pengjiayu	Remote	122.08	25.63	2010	1438	8.9	10.2	[26]
Mt. Waliguan	Remote	100.898	36.287	2012–2014	290	6.9	2.0	[8]
Mt. Leigong	Remote	108.203	26.387	2008–2009	1533	4.0	6.1	[8]
Mt. Ailao	Remote	101.107	24.533	2011–2014	1931	3.7	7.2	[8]
Mt. Damei	Remote	121.565	29.632	2012–2014	1621	3.7	6.0	[8]
Mt. Changbai	Remote	128.112	42.403	2011–2014	751	7.4	5.6	[8]
Bayinbuluk	Remote	83.717	42.893	2013–2014	266	7.7	2.0	[8]
Lhasa	Urban	91.12	29.63	2010	359	24.9	8.2	[27]
Guiyang	Urban	106.724	26.573	2012–2013	1057	11.9	12.6	[8]
Chongqing	Rural	106.58	29.52	2010–2014	1104	34.3	37.83	[28]
Seoul, Korea	Rural	127	37.51	2006–2007	1235–1645	10.1–16.3	16.8–20.2	[29]
10 sites around Japan	-	-	-	2004–2015	583–2147	-	5.8–18.0	[30]
17 sites of GMOS	Remote	-	-	2011–2015	47–1364	2.6–15.0	0.1–10	[9]

"-" means no data.

The annual wet deposition during the sampling period ranged from 2.6 to 9.8 $\mu g\ m^{-2}\ yr^{-1}$, with a multi-year annual average of 4.9 $\mu g\ m^{-2}\ yr^{-1}$. The Hg wet deposition flux in Chongming was lower than that at tropical background stations, such as Pengjiayu (10.2 $\mu g\ m^{-2}\ yr^{-1}$), Puerto Rico (27.9 $\mu g\ m^{-2}\ yr^{-1}$), Mt. Lulin (32.3 $\mu g\ m^{-2}\ yr^{-1}$), and Mt. Ailao (7.2 $\mu g\ m^{-2}\ yr^{-1}$), and higher than that at temperate background stations such as Mt. Waliguan (2.0 $\mu g\ m^{-2}\ yr^{-1}$), Pallas (1.2 $\mu g\ m^{-2}\ yr^{-1}$), Ny-Alesund (2.5 $\mu g\ m^{-2}\ yr^{-1}$), and Bayinbuluk (2.0 $\mu g\ m^{-2}\ yr^{-1}$) (Table 1). The rainfall depths were thought to be the main drivers of the different Hg wet deposition flux amounts [8]. The rainfall amount differences were attributed to variations of latitude and temperate regions have lower rainfall amounts than the subtropical and tropical regions (Figure S3). In addition, the Hg wet deposition flux amount in Chongming was lower than that at the urban areas in the YRD, such as Nanjing (56.5 µg m^{-2} yr^{-1}) and Shanghai (304 µg m^{-2} yr^{-1}) [14]. The VWM concentration in Chongming was lower than that at Nanjing and Shanghai (10.1–30.7 ng L^{-1}) and the annual rainfall depth in Chongming (834 mm) was lower than that of the downtown area of Shanghai (1100–1400 mm) due to the urban heat island effect [31]. The lower rainfall amounts and VWM Hg concentration both contributed to relatively low Hg wet deposition flux amounts in Chongming. Compared to the Hg wet deposition measurement in 2008–2009 in Chongming, the VWM Hg concentration showed a significant reduction from 62 ng L^{-1} in 2008–2009 to 7.6 ng L^{-1} in 2014–2018 [24]. The reduction of rainwater Hg concentrations suggests the atmospheric Hg concentrations declined in recent decades in Chongming, which is similar to the long-term variation of anthropogenic atmospheric Hg emissions in China [15,32].

Seasonal Hg wet deposition flux reached a maximum in summer and a minimum during winter, a pattern that is similar to the seasonal rainfall variation (Figure 2). Such seasonal patterns have been observed in eastern China, such as at Mt. Changbai, Qingdao, and Mt. Damei [8,28]. The air mass is transported from the Pacific Ocean to eastern China under the impact of the Asian summer monsoon with a significant amount of rainfall during summer [33]. The enhanced rainfall contributes to GEM oxidation and GOM dissolution in cloud water, leading to elevated Hg wet deposition flux during summer [34–36]. The VWM Hg concentrations were at minimum values in July and at maximum values in January. Anthropogenic emissions and rainfall amounts were the dominant factors in the VWM Hg concentrations. The anthropogenic atmospheric Hg emissions increased in the YRD region due to enhanced energy consumption during the winter [37]. The high PM$_{2.5}$ concentrations resulted in more GEM oxidation with higher GOM/PBM concentrations [19,38]. Meanwhile, airmass from northern China and Mongolia resulted in cold and dry weather with less rainfall in Chongming during winter. The enhanced PBM and lower rainfall amounts contributed to elevated VWM Hg concentrations during the winter.

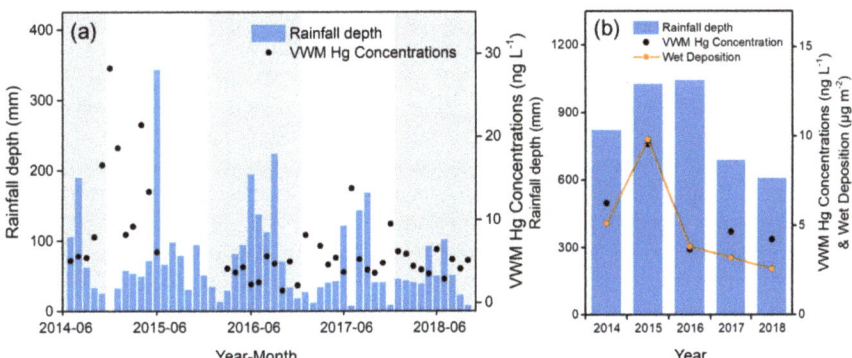

Figure 2. Time series of (**a**) monthly and (**b**) annual precipitation amounts and VWM Hg concentrations from June 2014 to October 2018. The alternate grey-white blocks represent different years. The missing black dots from June 2015 to February 2016 were due to instrument problems.

3.2. Impacts of Anthropogenic Emissions on Hg Wet Deposition Flux

To identify the major source regions of Hg in rainfall and investigate the influence of anthropogenic emissions, 48 h backward trajectories of 251 rainfall events were simulated during the sampling period. Chongming Island is affected by a subtropical monsoon climate, with 80% percent of rainfall occurring during the warm season (April–September). Thus, the 502 backward trajectories were divided into two parts by warm (April–September) and cold (October–March) seasons (Figure 2). During the warm season, cluster 1, cluster 2, and cluster 3 showed the airmass pathway from the Pacific Ocean, contributing to 65% of the rainfall events. The airmass from the Pacific Ocean during the warm season (clusters 1,2,3) contained a large amount of rainwater (13.92 mm) with relatively low VWM Hg concentrations (5.83 ng L^{-1}). Cluster 4 originated from northern China and contained less rainfall (9.38 mm) with enhanced VWM Hg concentration (7.91 ng L^{-1}). Figure 3b shows backward trajectories during the cold season. Clusters 2, 3, and 4 mainly originated from northern China and the Mongolia Plateau with 6.63 mm rainfall and 7.5 ng L^{-1} Hg concentration in airmass. The airmass from industrial area, such as the north China plain region and YRD region, could impact Hg wet deposition by scavenging atmospheric Hg into rainwater. Given that Hg wet deposition in Chongming was impacted by the airmass from eastern and northern China, the multi-year wet flux observation could reflect the effect of anthropogenic emission reduction in these areas.

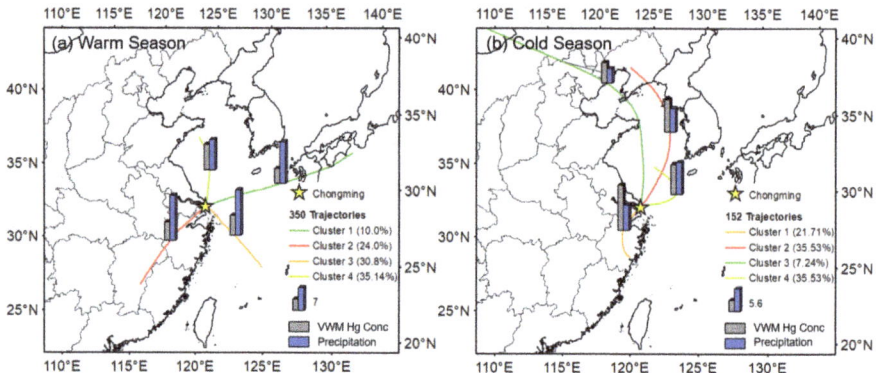

Figure 3. Forty-eight hour backward trajectories during the warm season (April–September) and cold season (October–March) in Chongming. The bar plot shows the mean rainfall depth and VWM Hg concentrations during the warm and cold season.

During the sampling period, the annual Hg wet deposition flux decreased from 8.6 µg m^{-2} in 2014–2015 to 3.8 µg m^{-2} in 2016. From June 2014 to December 2016, the decreasing trend (Sen's slope) of the VWM Hg concentration was 4.09 ng L^{-1} yr^{-1} (42% yr^{-1}) (Figure 4). In East Asia, previous long-term observations have mainly focused on atmospheric Hg concentrations instead of Hg wet deposition flux. Recent atmospheric Hg observation studies have also shown significant reductions in South Korea and Japan [39,40]. The decreasing trend of TGM in Chongming was 29.8% yr^{-1} from 2014 to 2016, which is similar to the significant reduction in the VWM Hg concentration during the sampling period [22] (Figure 4). The PBM decreased from 24.51 pg m^{-3} in 2014 to 19.79 pg m^{-3} in 2018, whereas GOM increased from 15.41 pg m^{-3} in 2014 to 20.08 pg m^{-3} in 2018 (Figure S4). The slight variation (39.86–41.05 pg m^{-3}) of reactive mercury (GOM+PBM) demonstrated the importance of GEM oxidation and dissolution, which is a dominant rainwater Hg source in the marine boundary layer due to GEM oxidation via bromine [34,36,41]. The VWM Hg reduction during the cold season (from 12.4 ng L^{-1} in 2014 to 4.6 ng L^{-1} in 2018) was more pronounced than the reduction in the warm season (from 5.4 ng L^{-1} in 2014 to 2.7 ng L^{-1} in 2018). After considering the rainout effect of VWM Hg concentration (Figure S5), the seasonal variation of Hg concentration suggests that the airmass from northern China had greater VWM Hg reductions than the airmass from overseas [16,42]. Many previous observations and modeling results have shown that the TGM/GEM reduction in China can be explained by reduced anthropogenic emissions in China [16,22,42]. Hg emissions decreased more quickly in the winter due to the changing of residential heating methods [16]. Therefore, the decreasing VWM Hg concentrations in Chongming between 2014 and 2016 were caused by atmospheric Hg concentration reductions, which could be attributed to the decline in Chinese anthropogenic Hg emissions in recent years [12].

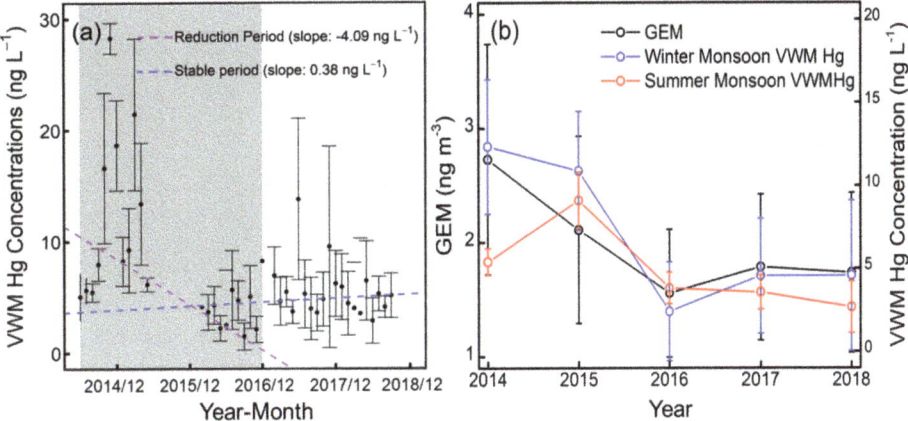

Figure 4. (a) The decreasing trend (Sen's slope) of the VWM Hg concentrations from June 2014 to December 2016 (reduction period, in grey panel) and from March 2016 to October 2018 (stable period, in white panel). (b) Comparison of gaseous elemental mercury (GEM) and the annual seasonal trend of VWM Hg from 2014 to 2018. The error bars indicate the standard error of the calculated monthly and annual VWM Hg concentrations and GEM.

From 2016 to 2018, the GEM and VWM Hg concentrations both remained at a relatively stable level with a slight increase of 0.38 ng L^{-1} (Sen's slope) in VWM Hg concentration. The average VWM Hg concentration during 2016–2018 (4.0±1.7 ng L^{-1}) was lower than that of the most subtropical remote background sites in east Asia (3.7–9.2 ng L^{-1}), whereas the TGM (1.7±0.1 ng m^{-3}) was only slightly higher than that of the background concentrations in the northern hemisphere (1.3–1.5 ng m^{-3}) [43]. The estimated annual Hg wet deposition flux further decreased from 2016 to 2018 under stable VWM Hg concentrations, which could be attributed to the interannual rainfall variation.

3.3. Impact of Meteorological Conditions on Hg Wet Deposition

Meteorological conditions have been found to be an important influential factor on Hg wet deposition flux [25,44,45]. A significant positive correlation between the daily rainfall depth and Hg wet deposition flux was observed (Figure 5, $R^2 = 0.55$, $p < 0.01$). This explained variation was slightly lower than that of the previous studies in temperate and subtropical remote regions (R^2 ranging from 0.6 to 0.8), which may be attributed to the influence of anthropogenic emissions [8,25,30,46]. Furthermore, the relationship between the VWM Hg concentration and wet deposition was not pronounced ($R^2 = 0.05$), which highlights the importance of rainfall amounts on Hg wet deposition flux in Chongming.

A pronounced rainout effect was observed in Chongming, where the VWM Hg decreased with the increased rainfall amounts (Figure 5). Rainfall type is thought to be an important influential factor on rainwater Hg concentration and to affect the rainout effect of rainwater mercury [35,45,47]. Convective rain can enhance GOM and PBM solutions, leading to more Hg wet deposition compared to non-convective rain [35,45,47]. Under the control of a cold high-pressure system, almost no convective rain occurs in October–March in Chongming. Nearly all of the convective weather occurs during April–September, about 15 times per year [48]. The correlation between VWM Hg concentration and rainfall depth showed little difference between the warm season ($R^2 = 0.04$, $p = 0.012$) and cold season ($R^2 = 0.05$, $p = 0.10$), suggesting the rain type was not an important factor in Chongming (Figure S5).

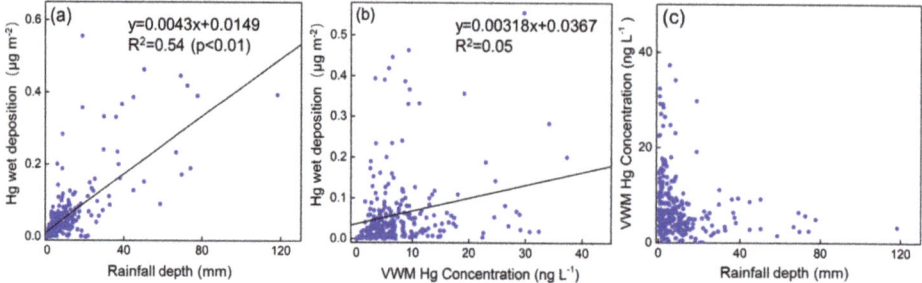

Figure 5. The correlation between (**a**) rainfall depth and (**b**) VWM Hg concentration and wet deposition of Hg. The fitted line was calculated by the least-squares method. (**c**) The scatter plot between rainfall depth and VWM Hg concentrations.

Between 2016 and 2018, the Hg wet deposition flux decreased from 3.8 to 2.5 μg m^{-2} with a stable VWM Hg concentration. The annual rainfall depth decreased from 1027 mm in 2014 to 608 mm in 2018, with the variation in the warm season being more pronounced (845–356 mm) than that in the cold season (163–320 mm) (Figure 6). The reduction of rainfall depth in the warm season from 845 mm in 2016 to 356 mm in 2018 dominated the Hg wet deposition flux reduction. The interannual rainfall depth variation in the warm season reflected the variation of large-scale circulation, such as monsoon activity [49]. The East Asian Summer Monsoon Index (EASMI) demonstrates the dominant wind variation from winter to summer at 850 hPa in East Asia and has been widely used in previous meteorological studies [49,50]. When the EASMI <0, it means the dominant wind direction during summer was reversed compare to the wind direction during winter. When the EASMI >0, it means the dominant wind direction during summer was similar to the wind direction during winter. The EASMI has a good correlation (R^2 = 0.40, p = 0.25, Figure S6) with the warm season rainfall in Chongming, and is comparable with the previous meteorological studies in the YRD region (R^2 = 0.3~0.5) [50]. The EASMI verified the interannual variation of rainfall amounts in the warm season during the sampling period. Thus, the large-scale meteorological circulation could impact wet deposition in Chongming by increasing or decreasing rainfall depth, and thus lead to the Hg wet deposition reduction from 2016 to 2018.

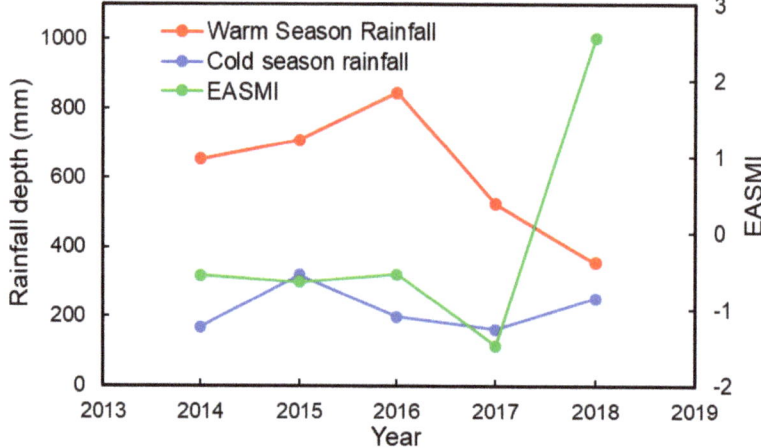

Figure 6. The annual variation in rainfall depth of warm and cold seasons and the East Asian Summer Monsoon Index (EASMI) index of the warm season during the sampling period.

4. Conclusions

In this study, multi-year daily rainfall events were collected and measured in Chongming from 2014 to 2018. During the sampling period, the mean VWM Hg concentration (7.6 ng L^{-1}) in Chongming was within the range of temperate and subtropical background observation sites in East Asia (3.7–9.2 ng L^{-1}) and lower than the urban site of YRD region (10.1–30.7 ng L^{-1}), reflecting the relatively clean environment in Chongming. The annual Hg wet deposition flux ranged from 2.6 to 9.8 µg m^{-2} yr^{-1} with a mean value of 4.9 µg m^{-2} yr^{-1}. The seasonal variation of Hg wet deposition flux was characterized by a maximum in June and a minimum in December, induced by the variation of rainfall amounts.

During the sampling period, a pronounced Hg wet reduction was observed from 8.6 µg m^{-2} in 2014–2015 to 2.6 µg m^{-2} in 2018, with an annual average of 4.9 µg m^{-2}. The Hg wet deposition reduction was attributed to the decreasing VWM Hg concentrations and reduction in the interannual rainfall. The Hg wet deposition flux decreased from 8.6 µg m^{-2} in 2014–2015 to 3.8 µg m^{-2} in 2016, with the VWM Hg concentration decreasing by 4.1 ng L^{-1} yr^{-1}. The reduced VWM Hg concentrations were driven by anthropogenic Hg emissions reductions, which were verified by the reduced atmospheric Hg concentrations. A further reduction in Hg wet deposition flux was observed under stable VWM Hg concentrations from 2016 to 2018. The interannual variation of the Asian summer monsoon activity resulted in the reduction in the rainfall from 845 mm in 2016 to 356 mm in 2018 during the warm season. These multi-year Hg wet deposition flux measurements could provide insights into the impact of Hg emissions reductions and interannual meteorological variations. Thus, we recommend using long-term Hg wet deposition flux values to evaluate the impact of anthropogenic emissions reductions and interannual meteorological conditions in future mercury assessment programs.

Supplementary Materials: The following are available online at http://www.mdpi.com/2073-4433/11/12/1301/s1, Figure S1: (a,b) Location of Chongming background observation station; (c) the photo of sampling weather station; Figure S2: Calibration plot for analysis of mercury by Tekran 2600; Figure S3: The (a) latitude (b) altitude (c) rainfall amount vs. wet Hg deposition in selected monitoring station of northern hemisphere; Figure S4: The annual GOM and PBM concentration in Chongming from 2014 to 2018. The Tekran 2537/1130/1135 encounter operation problem in 2015 and 2017, making it difficult to use annual mean concentration directly; Figure S5: Correlation between VWM Hg concentration and rainfall depth during (a) warm season and (b) cold season; Figure S6. Correlation of warm season rainfall depth and East Asia Summer Monsoon Index during sampling period.

Author Contributions: Y.T. and Q.W. designed the research. Y.T., W.G. and J.L. performed the research. Y.T. analyzed the data. Y.T., Q.W., S.W., K.L., Z.L. and D.H. wrote the manuscript with contributions from all other co-authors. All authors have read and agreed to the published version of the manuscript.

Funding: This research was funded by National Natural Science Foundation of China (21625701) and the Samsung Advanced Institute of Technology. Youth Program of National Natural Science Foundation of China: Grant No. 42007190. National Postdoctoral innovative talent program: Grant No. BX20190169; Tsinghua University "Shuimu Scholar" project: No. 2019SM061; China Postdoctoral Science Foundation Funded Project, Grant No. 2019M660672. National Key R&D Program of China (No. 2016YFC0201900). Dr. Shuxiao Wang acknowledges the support from the Tencent Foundation through the XPLORER PRIZE.

Conflicts of Interest: The authors declare no conflict of interest.

References

1. Lin, C.J.; Pehkonen, S.O. The chemistry of atmospheric mercury: A review. *Atmos. Environ.* **1999**, 2067–2079. [CrossRef]
2. Schroeder, W.H.; Munthe, J. Atmospheric mercury—An overview. *Atmos. Environ.* **1998**, *32*, 809–822. [CrossRef]
3. Lindberg, S.; Bullock, R.; Ebinghaus, R.; Engstrom, D.; Feng, X.; Fitzgerald, W.; Pirrone, N.; Prestbo, E.; Seigneur, C. A synthesis of progress and uncertainties in attributing the sources of mercury in deposition. *Ambio* **2007**, *36*, 19. [CrossRef]
4. Amos, H.M.; Jacob, D.J.; Holmes, C.D.; Fisher, J.A.; Wang, Q.; Yantosca, R.M.; Corbitt, E.S.; Galarneau, E.; Rutter, A.P.; Gustin, M.S.; et al. Gas-particle partitioning of atmospheric Hg(II) and its effect on global mercury deposition. *Atmos. Chem. Phys.* **2012**, *12*, 591–603. [CrossRef]

5. Gustin, M.S.; Huang, J.; Miller, M.B.; Peterson, C.; Jaffe, D.A.; Ambrose, J.; Finley, B.D.; Lyman, S.N.; Call, K.; Talbot, R.; et al. Do we understand what the mercury speciation instruments are actually measuring? Results of RAMIX. *Environ. Sci. Technol.* **2013**, *47*, 7295–7306. [CrossRef]
6. Zhang, L.; Wu, Z.; Cheng, I.; Wright, L.P.; Olson, M.L.; Gay, D.A.; Risch, M.R.; Brooks, S.; Castro, M.S.; Conley, G.D.; et al. The Estimated Six-Year Mercury Dry Deposition Across North America. *Environ. Sci. Technol.* **2016**, *50*, 12864–12873. [CrossRef]
7. Zhang, L.; Lyman, S.; Mao, H.; Lin, C.-J.; Gay, D.A.; Wang, S.; Sexauer Gustin, M.; Feng, X.; Wania, F. A synthesis of research needs for improving the understanding of atmospheric mercury cycling. *Atmos. Chem. Phys.* **2017**, *17*, 9133–9144. [CrossRef]
8. Fu, X.; Yang, X.; Lang, X.; Zhou, J.; Zhang, H.; Yu, B.; Yan, H.; Lin, C.-J.; Feng, X. Atmospheric wet and litterfall mercury deposition at urban and rural sites in China. *Atmos. Chem. Phys.* **2016**, *16*, 11547–11562. [CrossRef]
9. Sprovieri, F.; Pirrone, N.; Bencardino, M.; D'Amore, F.; Angot, H.; Barbante, C.; Brunke, E.-G.; Arcega-Cabrera, F.; Cairns, W.; Comero, S.; et al. Five-year records of mercury wet deposition flux at GMOS sites in the Northern and Southern hemispheres. *Atmos. Chem. Phys.* **2017**, *17*, 2689–2708. [CrossRef]
10. UNEP. Global Mercury Assessment 2018. Available online: https://www.unenvironment.org/resources/publication/global-mercury-assessment-2018 (accessed on 30 November 2020).
11. Prestbo, E.M.; Gay, D.A. Wet deposition of mercury in the U.S. and Canada, 1996–2005: Results and analysis of the NADP mercury deposition network (MDN). *Atmos. Environ.* **2009**, *43*, 4223–4233. [CrossRef]
12. Zhang, Y.; Jacob, D.J.; Horowitz, H.M.; Chen, L.; Amos, H.M.; Krabbenhoft, D.P.; Slemr, F.; St Louis, V.L.; Sunderland, E.M. Observed decrease in atmospheric mercury explained by global decline in anthropogenic emissions. *Proc. Natl. Acad. Sci. USA* **2016**, *113*, 526. [CrossRef] [PubMed]
13. Horowitz, H.M.; Jacob, D.J.; Amos, H.M.; Streets, D.G.; Sunderland, E.M. Historical Mercury releases from commercial products: Global environmental implications. *Environ. Sci. Technol.* **2014**, *48*, 10242–10250. [CrossRef] [PubMed]
14. Zhu, J.L.; Wang, T.J.; Talbot, R.; Mao, H.T.; Yang, X.; Fu, C.; Sun, J.; Zhuang, B.; Li, S.; Han, Y.; et al. Characteristics of atmospheric mercury deposition and size-fractionated particulate mercury in urban Nanjing, China. *Atmos. Chem. Phys.* **2014**, *14*, 2233–2244. [CrossRef]
15. Wu, Q.R.; Wang, S.X.; Li, G.L.; Liang, S.; Lin, C.J.; Wang, Y.; Cai, S.Y.; Liu, K.Y.; Hao, J.M. Temporal Trend and Spatial Distribution of Speciated Atmospheric Mercury Emissions in China During 1978–2014. *Environ. Sci. Technol.* **2016**, *50*, 13428–13435. [CrossRef]
16. Liu, K.Y.; Wu, Q.R.; Wang, L.; Wang, S.X.; Liu, T.H.; Ding, D.; Tang, Y.; Li, G.L.; Tian, H.Z.; Duan, L.; et al. Measure-Specific Effectiveness of Air Pollution Control on China's Atmospheric Mercury Concentration and Deposition during 2013–2017. *Environ. Sci. Technol.* **2019**, *53*, 8938–8946. [CrossRef]
17. Wu, Q.; Li, G.; Wang, S.; Liu, K.; Hao, J. Mitigation options of atmospheric Hg emissions in China. *Environ. Sci. Technol.* **2018**, *52*, 12368–12375. [CrossRef]
18. National Bureau of Statiatics of China. China Statistical Yearbook. 2019. Available online: http://www.stats.gov.cn/tjsj/ndsj/2019/indexeh.htm (accessed on 30 November 2020).
19. Zhang, L.; Wang, L.; Wang, S.; Dou, H.; Li, J.; Li, S.; Hao, J. Characteristics and Sources of Speciated Atmospheric Mercury at a Coastal Site in the East China Sea Region. *Aerosol. Air Qual. Res.* **2017**, *17*, 2913–2923. [CrossRef]
20. Draxler, R.R.; Hess, G.D. An overview of the hysplit-4 modeling system for trajectories. *Aust. Meteorol. Mag.* **1998**, *47*, 295–308.
21. Dorling, S.R.; Davis, T.D.; Pierce, C.E. Cluster analysis: A technique for estimating the synoptic meteorological controls on air and precipitation chemistry—Method and applications. *Atmos. Environ.* **1992**, *26*, 2581–2585. [CrossRef]
22. Tang, Y.; Wang, S.; Wu, Q.; Liu, K.; Wang, L.; Li, S.; Gao, W.; Zhang, L.; Zheng, H.; Li, Z.; et al. Recent decrease trend of atmospheric mercury concentrations in East China: The influence of anthropogenic emissions. *Atmos. Chem. Phys.* **2018**, *18*, 8279–8291. [CrossRef]
23. Cole, A.S.; Steffen, A.; Pfaffhuber, K.A.; Berg, T.; Pilote, M.; Poissant, L.; Tordon, R.; Hung, H. Ten-year trends of atmospheric mercury in the high Arctic compared to Canadian sub-Arctic and mid-latitude sites. *Atmos. Chem. Phys.* **2013**, *13*, 1535–1545. [CrossRef]

24. Shi, G.; Chen, Z.; Teng, J.; Li, Y. Spatio-temporal variation of total mercury in precipitation in the largest industrial base in China: Impacts of meteorological factors and anthropogenic activities. *Tellus B Chem. Phys. Meteorol.* **2015**, *67*, 25660. [CrossRef]
25. Nguyen, L.S.P.; Sheu, G.-R. Four-year Measurements of Wet Mercury Deposition at a Tropical Mountain Site in Central Taiwan. *Aerosol Air Qual. Res.* **2019**, *19*, 2043–2055. [CrossRef]
26. Sheu, G.-R.; Lin, N.-H. Characterizations of wet mercury deposition to a remote islet (Pengjiayu) in the subtropical Northwest Pacific Ocean. *Atmos. Environ.* **2013**, *77*, 474–481. [CrossRef]
27. Huang, J.; Kang, S.; Wang, S.; Wang, L.; Zhang, Q.; Guo, J.; Wang, K.; Zhang, G.; Tripathee, L. Wet deposition of mercury at Lhasa, the capital city of Tibet. *Sci. Total Environ.* **2013**, *447*, 123–132. [CrossRef]
28. Chen, L.; Li, Y.; Liu, C.; Guo, L.; Wang, X. Wet deposition of mercury in Qingdao, a coastal urban city in China: Concentrations, fluxes, and influencing factors. *Atmos. Environ.* **2018**, *174*, 204–213. [CrossRef]
29. Seo, Y.-S.; Han, Y.-J.; Choi, H.-D.; Holsen, T.M.; Yi, S.-M. Characteristics of total mercury (TM) wet deposition: Scavenging of atmospheric mercury species. *Atmos. Environ.* **2012**, *49*, 69–76. [CrossRef]
30. Sakata, M.; Marumoto, K. Wet and dry deposition fluxes of mercury in Japan. *Atmos. Environ.* **2005**, *39*, 3139–3146. [CrossRef]
31. Voogt, J.A.; Oke, T.R. Thermal remote sensing of urban climates. *Remote Sens. Environ.* **2003**, *86*, 370–384. [CrossRef]
32. Zhang, L.; Wang, S.X.; Wang, L.; Wu, Y.; Duan, L.; Wu, Q.R.; Wang, F.Y.; Yang, M.; Yang, H.; Hao, J.M.; et al. Updated emission inventories for speciated atmospheric mercury from anthropogenic sources in China. *Environ. Sci. Technol.* **2015**, *49*, 3185–3194. [CrossRef]
33. Fu, X.W.; Zhang, H.; Yu, B.; Wang, X.; Lin, C.J.; Feng, X.B. Observations of atmospheric mercury in China: A critical review. *Atmos. Chem. Phys.* **2015**, *15*, 9455–9476. [CrossRef]
34. Horowitz, H.M.; Jacob, D.J.; Zhang, Y.; Dibble, T.S.; Slemr, F.; Amos, H.M.; Schmidt, J.A.; Corbitt, E.S.; Marais, E.A.; Sunderland, E.M. A new mechanism for atmospheric mercury redox chemistry: Implications for the global mercury budget. *Atmos. Chem. Phys. Discuss.* **2017**, 1–33. [CrossRef]
35. Holmes, C.D.; Krishnamurthy, N.P.; Caffrey, J.M.; Landing, W.M.; Edgerton, E.S.; Knapp, K.R.; Nair, U.S. Thunderstorms Increase Mercury Wet Deposition. *Environ. Sci. Technol.* **2016**, *50*, 9343–9350. [CrossRef] [PubMed]
36. Holmes, C.D.; Jacob, D.J.; Corbitt, E.S.; Mao, J.; Yang, X.; Talbot, R.; Slemr, F. Global atmospheric model for mercury including oxidation by bromine atoms. *Atmos. Chem. Phys.* **2010**, *10*, 12037–12057. [CrossRef]
37. Liu, K.Y.; Wang, S.X.; Wu, Q.R.; Wang, L.; Ma, Q.; Zhang, L.; Li, G.L.; Tian, H.Z.; Duan, L.; Hao, J.M. A Highly Resolved Mercury Emission Inventory of Chinese Coal-Fired Power Plants. *Environ. Sci. Technol.* **2018**, *52*, 2400–2408. [CrossRef]
38. Hong, Y.W.; Chen, J.S.; Deng, J.J.; Tong, L.; Xu, L.L.; Niu, Z.C.; Yin, L.Q.; Chen, Y.T.; Hong, Z.Y. Pattern of atmospheric mercury speciation during episodes of elevated PM2.5 levels in a coastal city in the Yangtze River Delta, China. *Environ. Pollut.* **2016**, *218*, 259–268. [CrossRef]
39. Kim, K.-H.; Brown, R.J.C.; Kwon, E.; Kim, I.-S.; Sohn, J.-R. Atmospheric mercury at an urban station in Korea across three decades. *Atmos. Environ.* **2016**, *131*, 124–132. [CrossRef]
40. Marumoto, K.; Suzuki, N.; Shibata, Y.; Takeuchi, A.; Takami, A.; Fukuzaki, N.; Kawamoto, K.; Mizohata, A.; Kato, S.; Yamamoto, T.; et al. Long-Term Observation of Atmospheric Speciated Mercury during 2007–2018 at Cape Hedo, Okinawa, Japan. *Atmosphere* **2019**, *10*, 362. [CrossRef]
41. Wang, S.; McNamara, S.M.; Moore, C.W.; Obrist, D.; Steffen, A.; Shepson, P.B.; Staebler, R.M.; Raso, A.R.W.; Pratt, K.A. Direct detection of atmospheric atomic bromine leading to mercury and ozone depletion. *Proc. Natl. Acad. Sci. USA* **2019**, *116*, 14479–14484. [CrossRef]
42. Wu, Q.; Tang, Y.; Wang, S.; Li, L.; Deng, K.; Tang, G.; Liu, K.; Ding, D.; Zhang, H. Developing a statistical model to explain the observed decline of atmospheric mercury. *Atmos. Environ.* **2020**, *243*, 117868. [CrossRef]
43. Sprovieri, F.; Pirrone, N.; Bencardino, M.; D'Amore, F.; Carbone, F.; Cinnirella, S.; Mannarino, V.; Landis, M.; Ebinghaus, E.; Weigelt, A.; et al. Atmospheric mercury concentrations observed at ground-based monitoring sites globally distributed in the framework of the GMOS network. *Atmos. Chem. Phys.* **2016**, *16*, 11915–11935. [CrossRef] [PubMed]
44. Mao, H.; Ye, Z.; Driscoll, C. Meteorological effects on Hg wet deposition in a forested site in the Adirondack region of New York during 2000–2015. *Atmos. Environ.* **2017**, *168*, 90–100. [CrossRef]

45. Shanley, J.B.; Engle, M.A.; Scholl, M.; Krabbenhoft, D.P.; Brunette, R.; Olson, M.L.; Conroy, M.E. High Mercury Wet Deposition at a "Clean Air" Site in Puerto Rico. *Environ. Sci. Technol.* **2015**, *49*, 12474–12482. [CrossRef] [PubMed]
46. Nguyen, L.S.P.; Sheu, G.R.; Lin, D.W.; Lin, N.H. Temporal changes in atmospheric mercury concentrations at a background mountain site downwind of the East Asia continent in 2006–2016. *Sci. Total Environ.* **2019**, *686*, 1049–1056. [CrossRef]
47. Kaulfus, A.S.; Nair, U.; Holmes, C.D.; Landing, W.M. Mercury Wet Scavenging and Deposition Differences by Precipitation Type. *Environ. Sci. Technol.* **2017**, *51*, 2628–2634. [CrossRef]
48. Yang, L.H.; Yin, H.P.; Wang, H.; Tao, L. Statistical analysis of severe convective weather in shanghai area during period of recent 10 years. *Atmos. Sci. Res. Appl.* **2007**, *2*, 84–91.
49. Li, J.; Wu, Z.; Jiang, Z.; He, J. Can Global Warming Strengthen the East Asian Summer Monsoon? *J. Clim.* **2010**, *23*, 6696–6705. [CrossRef]
50. Li, J.P.; Zeng, Q.C. A new monsoon index and the geographical distribution of the global monsoons. *Adv. Atmos. Sci.* **2003**, *20*, 299–302.

Publisher's Note: MDPI stays neutral with regard to jurisdictional claims in published maps and institutional affiliations.

© 2020 by the authors. Licensee MDPI, Basel, Switzerland. This article is an open access article distributed under the terms and conditions of the Creative Commons Attribution (CC BY) license (http://creativecommons.org/licenses/by/4.0/).

Communication

A Collaborative Training Program to Assess Mercury Pollution from Gold Shops in Guyana's Artisanal and Small-Scale Gold Mining Sector

Samantha T. Brown [1], Lloyd L. Bandoo [2], Shenelle S. Agard [2], Shemeiza T. Thom [2], Tamara E. Gilhuys [2], Gautham K. Mudireddy [1], Arnith V. Eechampati [1], Kazi M. Hasan [1], Danielle C. Loving [1], Caryn S. Seney [1] and Adam M. Kiefer [1],*

1. Department of Chemistry, Mercer University, Macon, GA 31201, USA; samantha.brown@live.mercer.edu (S.T.B.); Gautham.K.Mudireddy@live.mercer.edu (G.K.M.); Arnith.Venugopal.Eechampati@live.mercer.edu (A.V.E.); Kazi.M.Hasan@live.mercer.edu (K.M.H.); danielle.cherrice.loving@live.mercer.edu (D.C.L.); SENEY_CS@mercer.edu (C.S.S.)
2. Guyana Geology and Mines Commission, Upper Brickdam, Georgetown, Guyana; lloyd_bandoo@ggmc.gov.gy (L.L.B.); shenelle_agard@ggmc.gov.gy (S.S.A.); shemeiza_thom@ggmc.gov.gy (S.T.T.); tamara_gilhuys@ggmc.gov.gy (T.E.G.)
* Correspondence: kiefer_am@mercer.edu

Received: 6 June 2020; Accepted: 30 June 2020; Published: 6 July 2020

Abstract: A three-phase, 11-day training program designed to monitor elemental mercury (Hg^0) emissions originating from gold shops was conducted in Georgetown and Bartica, Guyana, during May of 2019. The first phase consisted of interactive lectures and discussions on mercury use in artisanal and small-scale gold mining throughout Guyana, the region, and the world. In addition, specific training in the theory and use of analytical instrumentation to quantify Hg^0 pollution associated with the processing of amalgams and sponge gold occurred. Trainees participated in the mapping of smelting facilities in Georgetown where, outside of one gold shop, Hg^0 concentrations exceeded 100,000 ng/m^3. During the second phase of training, a subset of trainees traveled to Bartica, where they mapped the town center to identify point sources of Hg^0 pollution, all of which corresponded to the location of shops where amalgams and sponge gold were heated and purchased. Once mapped, Hg^0 concentrations were measured during the smelting of gold inside the Guyana Gold Board (GGB) facility and two privately-owned gold shops. Maximum Hg^0 concentrations at the GGB facility did not exceed 98,700 ng/m^3 during the measurement period, while maximum concentrations at the two privately owned shops were measured as 527,500 ng/m^3 and 302,200 ng/m^3. With guidance from the training team, trainees were responsible for the collection and interpretation of all data. The third phase of the training involved the collaborative production of a report summarizing the findings from the training. This work represents the first formal training opportunity for the assessment of Hg^0 concentrations in and around gold shops in Guyana, and provides baseline data to assist the government of Guyana to generate air quality standards for Hg^0 emissions.

Keywords: mercury; artisanal and small-scale gold mining; amalgam; Minamata Convention on Mercury; gold shop

1. Introduction

1.1. The Mercury Problem in Artisanal and Small-Scale Gold Mining

In 2014, it was estimated that 16 million artisanal and small-scale gold miners annually produced 380–450 tonnes (t) of gold, representing ≈17–20% of global gold production [1]. Although the estimated number of miners may change on the basis of global gold prices and other factors, it is undeniable that artisanal and small-scale gold mining (ASGM) plays an important role not only in global gold production, but also as a meaningful source of employment throughout the developing world [1–10]. In spite of its socioeconomic importance, ASGM can severely impact both environmental and human health through deforestation [11,12], habitat loss [13,14], social issues [15,16], and work-related injuries [17,18]. ASGM is also closely associated with mercury pollution derived from ore processing [19–27] and is now the leading source of anthropogenic mercury emissions on the planet [28,29].

Although ASGM practices are diverse and dependent upon the ore being processed, the technologies available, and the region in which mining is occurring, the majority of ASGM is conducted using elemental mercury (Hg^0) at some stage of the process. Amalgamation with Hg^0 concentrates gold, separating it from unwanted minerals. Subsequent heating of the amalgam evaporates the Hg^0, revealing the sponge gold that can then be sold. During these processes, Hg^0 is lost to both the tailings and the atmosphere. Mercury is used because it is inexpensive, readily available, requires no special training, acts quickly, and can be used independently. Miners often believe that the money they earn from mining outweighs the health risks associated with the Hg^0 vapor—an acute and chronic toxin [30–33]. This issue is inadvertently supported by the fact that Hg^0 vapor is not detectable by the human senses, and physiological changes associated with chronic mercury poisoning develop slowly, leading many miners to believe that the health effects of mercury are not severe. Further, the behavior of Hg^0 in the environment compounds the threat to human health. Hg^0 is both a mobile and persistent pollutant; once Hg^0 enters the global mercury cycle, it can be converted into inorganic mercury (Hg^{2+}) and ultimately methyl mercury (CH_3Hg^+), which is readily bioaccumulated and biomagnified [34,35].

1.2. ASGM in Guyana

ASGM in Guyana has a long history, largely beginning with a gold rush in the mid 19th century that led to people leaving agricultural work on the coast for mining activities in the interior [36]. Guyanese miners, locally known as pork-knockers for their habit of carrying large barrels of dried pork into the interior for sustenance, were initially forced to follow bodies of water and only mine near-surface gold due to the inaccessibility of the jungle. As the transport of mechanized equipment inland has become more manageable, miners adopted new technologies that allowed them to access new deposits while improving throughput of material. While river dredging and hard-rock mining both occur in Guyana, the majority of ore is extracted using land dredges.

Ore extraction and processing is relatively uniform across Guyana and employs hydraulic mining. The mining claim is cleared, and the overburden is removed using heavy equipment. A mining team then uses water monitors to wash gold-containing soil, clay, sediment, etc. to the bottom of the pit, where the resulting slurry is subsequently pumped out of the pit to be concentrated on a sluice box. The vast majority of mining operations concentrate gold on carpet material that lines the sluice box. After an appropriate amount of material has been collected, the miners "wash down" and amalgamate the concentrate. The Hg^0-contaminated tailings are discarded, and the amalgam is heated on site, revealing the sponge gold. Miners bring the sponge gold to a business that purchases gold, where it is reheated in an attempt to drive off residual Hg^0 prior to sale.

All gold extracted in Guyana is required to be sold to the state, and as such miners bring their sponge gold to either one of the Guyana Gold Board (GGB) locations in Georgetown or Bartica, or to one of the eight private gold dealers licensed by the GGB. A gold dealer is licensed to buy on behalf of the GGB and export. Miners may also bring their sponge gold to one of several gold traders licensed by the GGMC that can purchase gold and resell it within Guyana [37,38]. Although Guyanese miners can easily smelt and sell gold to licensed dealers and traders, some miners choose to sell to unlicensed gold buyers. These private shops may offer a higher price for gold than the GGB, or may provide miners with other incentives such as supplies or the ability to work the shopkeepers claim [39]. Although all gold buyers are required by law to sell the gold to the GGB, there is an ongoing issue with unlicensed gold buyers illegally smuggling gold out of Guyana. In 2016, the Minister of Natural Resources estimated that ≈15,000 ounces (Ozs) of gold were smuggled out of the country each week [40–42].

In this report, the term "gold shop" will refer to any private business that purchases gold, legally or illegally. Gold shops are not unique to Guyana and are found in many Central and South American countries with ASGM. These shops are often located in residential and populated business areas, and are known point sources for Hg^0 pollution [43–47]. Unlike GGB facilities that have adequate ventilation systems to remove adventitious Hg^0 during the smelting process, the vast majority of gold shops do not. These shops not only reheat or smelt sponge gold, but miners sometimes burn entire amalgams, releasing large quantities of Hg^0 vapor into the shop and immediate surroundings. During burning, Hg^0 concentrations in the air often exceed levels that represent an immediate threat to human health [48,49].

1.3. ASGM and the Minamata Convention on Mercury

The Minamata Convention on Mercury is an international treaty designed specifically to decrease the use of Hg^0 and Hg-containing compounds, thus reducing anthropogenic emissions of the toxic metal [29,50,51]. Due to the sheer magnitude of emissions related to ASGM, Article 7 and the corresponding Annex C of the convention requires nations to reduce and, where feasible, eliminate both the use of Hg^0 and Hg^0 emissions that result from mining and processing activities. The treaty requires each signatory nation with ASGM to produce a detailed national action plan (NAP) highlighting efforts to curtail Hg^0 use. Countries are specifically tasked with setting their own goals and reduction targets, while developing a baseline estimate of the amount of Hg^0 used in ASGM activities. The NAP identifies actions to eliminate worst practices such as whole-ore amalgamation, cyanidation of Hg-contaminated tailings, and burning of amalgams in residential areas. In addition, nations are required to outline steps to formalize mining, and develop a public health strategy to address Hg^0 exposure in communities, particularly targeting vulnerable populations. While continued monitoring of Hg^0 emissions from ASGM activities is not specifically required after the filing of the NAP, reports on progress towards the nation's goals are required every three years.

Hg^0 use is only one component of a complex and varied mining culture, and efforts to limit Hg pollution are occasionally equated with attempts to eliminate ASGM, and thus the livelihood of miners. Successful implementation of the Minamata Convention will require experts from the social and natural sciences, education, engineering, health professions, and private business to work together to find solutions to complex issues arising at the interface of the environment and society [4,50,52–58]. A recent systematic review highlighted the importance of both education and collaboration in addressing Hg-related issues in ASGM communities [58]. The convention recognizes this as well, and there are numerous mechanisms for collaboration between signatory nations and other parties.

1.4. Training Program for Monitoring of Hg^0 in Air in ASGM Communities

A recent collaboration between the Guyana Geology and Mines Commission (GGMC) and Mercer University identified the need to further develop governmental competencies in both the collection and interpretation of scientific data in the field. The data collected and interpreted by the GGMC could be used both in the development of Guyana's NAP and subsequent assessment of progress towards meeting national goals. The collaboration relies on the experience of the GGMC in addressing the technical aspects and environmental ramifications of Guyanese mining, and Mercer's experience in conducting environmental analyses of Hg^0 emissions to air in ASGM communities. The GGMC, particularly the Mineral Processing Unit and the Environmental Division, have extensive experience in field work, mining camp safety assessment, and data collection. While the GGMC has instrumentation that allows them to monitor Hg^0 emissions in the field, they recognize that newer methods and instrumentation are available that would aid in both environmental assessment and limiting exposure of GGMC workers to Hg^0 contamination while conducting this assessment.

Monitoring Hg^0 at mining sites in Guyana is complicated; the vast amount of mining consists of concentrate amalgamation, with the amalgam being burned at the mining site only once per week. Oftentimes, the burning of the amalgam is conducted in private to avoid theft and potential violence as a result of disclosing the location of productive, remote mining sites. Measuring Hg^0 vapor in these environments is also complicated by the fact that many of these sites are deforested and open, leading to the rapid dispersal of Hg^0 in the atmosphere. However, the reheating of sponge gold and the smelting of gold at gold shops in populated, urban environments generates high concentrations of Hg^0 that can be measured with more accuracy [43,46]. As the open burning of amalgams and sponge gold, particularly burning activities in residential areas, are considered "actions to eliminate" under the Minamata Convention, it was decided that initial training would be conducted in ASGM communities with a number of gold shops. The monitoring of these shops is under the purview of the GGMC, and as such, it was decided that training using multiple portable atomic spectrometers for the detection of Hg^0 in the atmosphere would benefit the GGMC. Additional hand-held X-ray fluorescence analyzer training to determine heavy metal concentrations in soils and tailings would also occur. Herein, we present an overview of the training that took place, as well as the findings of the environmental assessment resulting from the training.

2. Materials and Methods

Soils and tailings were screened using an Olympus Vanta C Handheld X-ray fluorescence analyzer (XRF) with a field stand kit. Samples were analyzed using the manufacturer's GeoChem(2) method. The GeoChem(2) method utilizes a fundamental parameters calculation method [59–63]. Mercury concentrations in air were determined using four commercially available atomic absorption spectrometers: one Mercury Instruments Mercury Tracker 3000 IP (MTIP), one Mercury Instruments VM-3000 (VM-3000), and two Lumex RA-915 M (Lumex) spectrometers. The MTIP and VM-3000, calibrated by the manufacturer, measure ranges of 0–2,000,000 ng/m^3, have sensitivities of 0.1 $\mu g/m^3$, and have response times of 1 s. The Lumex employs Zeeman correction and has a significantly lower detection limit (0.5 ng/m^3). Both Lumex spectrometers were calibrated by the manufacturer prior to use. Areas with concentrations exceeding 50,000 ng/m^3 were actively avoided due to the instrument's calibration limits [64]. Concentrations occasionally exceeded 50,000 ng/m^3 due to shifting winds or unexpected activity within gold shops.

Concentrations of Hg^0 within gold shops and ventilated exhaust from gold shops on the streets and sidewalks were determined exclusively with the MTIP. When concentrations exceeded 1,500,000 ng/m^3, the MTIP was removed to an area of low concentrations of Hg^0 (<50 ng/m^3) and operated until measured concentrations decreased to less than 1000 ng/m^3.

All maps were generated from data collected by the Lumex less than 42,307 ng/m^3 to ensure that measurements remained on the calibration curves of both instruments. The mapping protocol has been previously described in the literature [65]. The Lumex was operated using the manufacturer's RAPID software and synced with a Garmin Oregon GPS unit. Both the Lumex and the GPS unit collected a sample every second. Upon the completion of data collection, all data were imported into Microsoft Excel, and the position was linked to concentration via time. Occasionally, a data point was generated with only position or concentration; these data were eliminated. Maximum Hg^0 values assigned to each unique set of coordinates were mapped.

Concentrations of Hg^0 in and around gold shops routinely exceeded the detection limit of the Lumex instrument. At high concentrations of Hg^0, the Lumex detector can become saturated, leading to the measurement of concentrations as less than 0 ng/m^3 until the re-establishment of a normalized baseline occurs. To avoid this, the MTIP was paired with the Lumex mapping teams to avoid areas where concentrations exceeded the detection limit. Any negative concentrations measured during monitoring were not mapped.

During training, heat maps were initially produced using Google Fusion Tables to rapidly plot the data; however, the heat map functionality of Google Fusion Tables was retired in December 2019. To that end, training and maps in this manuscript were generated using QGIS ("QGIS Development Team (2020). QGIS Geographic Information System. Open Source Geospatial Foundation Project. http://qgis.osgeo.org").

3. Results and Discussion

3.1. Overview of Training

Training was conducted 15–30 May 2019 in Georgetown and Bartica, Guyana. The training was divided into three phases, informally referred to as "discuss, do, and disseminate" (Figure 1). At the request of the GGMC, the initial "discuss" phase took place in Georgetown and included 47 participants from government, academia, non-governmental organizations (NGOs), and miners' organizations. The goal was to provide an overview of the project in addition to a foundation for subsequent training on instrumentation and data interpretation. Training in the field during this phase was intended for participants to learn how each instrument operated and could be used to generate maps depicting Hg^0 vapor concentrations associated with gold shops in Georgetown.

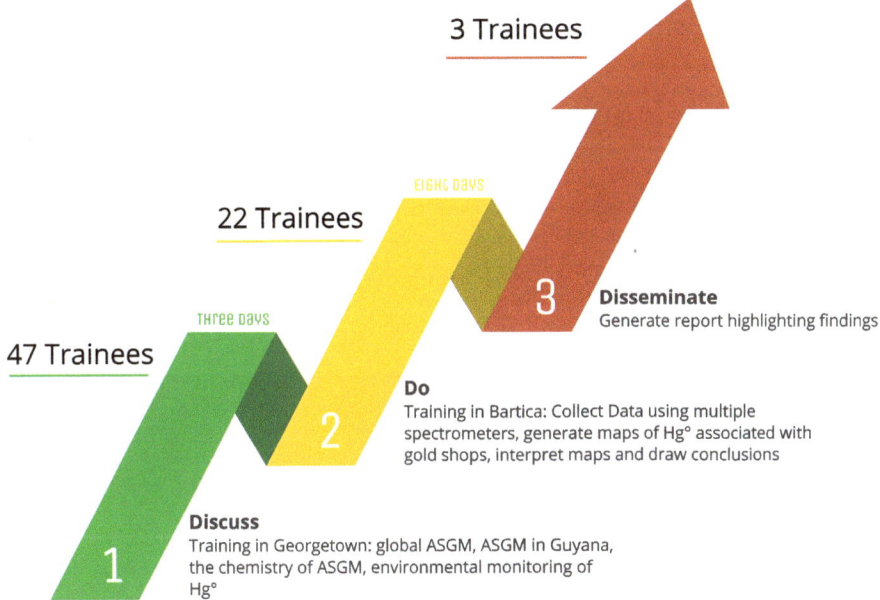

Figure 1. Overview of content and training activities.

The second "do" phase served as an eight-day training session in Bartica, where 22 select trainees from phase one were further trained to collect data using the Lumex and MTIP, map the data, and interpret the maps. Additional monitoring of emissions from gold shops was also conducted.

Finally, the third and final "disseminate" phase consisted of work done after completion of the formal training. Representatives from the GGMC were enlisted to formalize the findings of the training in a final report. The report was generated and reviewed by both collaborating institutions, with the goal of incorporating the peer-reviewed literature to support the justification of the project, the results of the training, and conclusions drawn from the results.

3.2. Phase I: Training in Georgetown

The "discuss" component of training was conducted 15–17 May in Georgetown, Guyana. Participants were selected by the GGMC and included representatives from the Environmental Division and Mineral Processing Unit of the GGMC, the Guyana Environmental Protection Agency (EPA), the Ministry of Natural Resources, the Ministry of Public Health, the University of Guyana, and the Guyana Gold Board. Additional representatives from the National Mining Syndicate, the Guyana

Gold and Diamond Miners Association, and the Guyana Women Miners' Organization were present, as were representatives from Conservation International Guyana and other NGOs.

The first day was dedicated to a non-scientific project "kickoff". Representatives from Mercer University, the Ministry of National Resources, and the GGMC provided a project overview to the participants, as well as the role of each organization in the project. The remainder of the first day was dedicated to an introductory lecture discussing global ASGM practices, Guyanese ASGM practices, Hg^0 use in ASGM, and an overview of the Minamata Convention on Mercury. A second lecture was then given, discussing the environmental fate of Hg^0 from ASGM including the global mercury cycle and the health effects of Hg^0. After the second lecture, a group discussion was moderated that engaged participants to offer their own observations on Hg^0 use in Guyana, particularly as to how it related to the mining process. During this conversation, chemical hygiene and safe storage of Hg^0 was discussed. It was unequivocally stated by both governmental representatives and miners alike that Hg^0 needed to be replaced in the ASGM process; however, currently there is no suitable replacement for Hg^0 in Guyana. This is not unique to Guyana. Hg^0 is inexpensive, readily available, requires no formal training for use, and can be used by an individual miner. Although mercury-free technologies and techniques have been developed for use in ASGM, none have been widely adopted [10,66–68].

The second day of training was more technical in nature, with detailed discussions related to the chemistry of artisanal and small-scale mining, mineral processing engineering, and analytical techniques for monitoring Hg^0 in the environment (Table 1). Because of the varied backgrounds of participants, care was taken to address each topic from an accessible perspective; while technical data was disseminated, it was placed in appropriate context for the audience, knowing that the second phase in Bartica would allow for additional technical training. Content was delivered through PowerPoint lectures, round-table discussions, and hands-on introduction to the instrumentation.

Table 1. Overview of topics covered on day two of the Georgetown training.

Topic	Content Delivery
Introduction to the chemical properties of Hg^0	Lecture
Mercury in the environment: Myths vs. Realities	Lecture, discussion
Mercury Pollution from ASGM: processing ore vs. processing amalgams	Lecture, discussion
Monitoring Hg^0 pollution in the environment	Lecture, discussion
Human health effects of Hg^0: safety in the field	Lecture, discussion
Techniques for soil analysis: XRF screening of metals in the field, laboratory analysis	Lecture, discussion, hands-on training
Techniques for monitoring Hg in the air: Hg^0 vs. total gaseous mercury (TGM)	Lecture, discussion, hands-on training
Case study on Peruvian ASGM mercury emissions: science vs. policy	Lecture, discussion
Introduction to operating principles of XRF and portable spectrometers	Lecture
Closing discussion: Hg^0 use in Guyana	Discussion

On day two, discussions were moderated on Hg^0 use from miners' perspectives. These included how the burning of amalgams in the field and in gold shops directly affect human and environmental health. The lectures and discussions were placed into the context of the Minamata Convention on Mercury, particularly the potential effects of the treaty on mercury use in Guyana in the future. A case study of a recent joint project between Mercer University and the Peruvian Ministerio del Ambiente (MINAM, Ministry of the Environment) was presented and discussed, including the results of mapping activities conducted in ASGM communities in the Peruvian Amazon. Hg^0 pollution originating from Peruvian gold shops was placed in the context of existing Peruvian air quality standards, which limit air concentrations to 2000 ng/m^3 of total gaseous mercury (TGM) over a 24 h period [69,70]. During the collaborative assessment of the Peruvian ASGM communities, Hg^0 concentrations exceeding 2,000,000 ng/m^3 were measured on the street outside these gold shops. As Hg^0 is a component of

TGM, the measured values clearly exceeded the legal limit for Hg^0 emissions to air by many orders of magnitude. However, because the technical norms at the time required that TGM be measured, the Hg^0 concentrations collected during the study period were inadequate to demonstrate that the air quality standards were exceeded. The work conducted by Mercer and MINAM led to the modification of Peruvian code to allow for the use of a conversion factor converting data collected by cold vapor atomic absorption spectroscopy (CV-AAS) with Zeeman correction into an estimate of TGM concentrations. Following the presentation of the case study, the current lack of a Guyanese Hg^0 emission standard was discussed with respect to the Peruvian case study.

To accommodate the large crowd, all four portable atomic absorption spectrometers and the handheld XRF were displayed. Trainees were split into multiple groups and allowed to handle each unit, examine how the unit was operated, and previewed the software, prior to rotating to the next instrument. For the VM-3000 and the MTIP, an amalgam (*parad shivling*, [71]) in a glass container was placed in front of each unit and opened so that trainees could see the concomitant increase of Hg^0 as measured by the instruments. Upon completing the demonstration, short clips of a YouTube video made at Mercer University allowing for the indirect visualization of Hg^0 vapor using a UV lamp and a thin-layer chromatography plate imbued with a fluorescent dye were shown to the participants [71,72]. On the basis of feedback and recommendations from the audience, especially representatives from the miners' organizations, we are currently modifying the video to directly engage the ASGM community.

Field Training in Georgetown

The final day of training was dedicated to using the MTIP and Lumex spectrometers to rapidly assess Hg^0 concentrations from gold shops in Georgetown. The trainees were split into two groups, one in the morning and one in the afternoon; both groups followed the same path through the city. As there has yet to be a comprehensive survey of gold shops in the city, this activity was largely based upon the knowledge of gold shops by GGMC employees. The Lumex spectrometers and MTIP were carried while walking through the city, and trainers demonstrated appropriate procedures and methods for data collection. Instruments were then handed to volunteer trainees, and they were assisted in collecting data by the trainers. During monitoring, two sites with elevated Hg^0 concentrations were located, both associated with gold shops (Figure 2). Concentrations of Hg^0 outside of both shops exceeded 10,000 ng/m^3, with one shop exceeding 100,000 ng/m^3 as measured by the MTIP.

The data were quickly processed and mapped using Google Heat Maps so that participants could see the sources of Hg^0 contamination. The quantitative heat map generated through QGIS can be found in Figure 2. While participants had already been instructed on how both chronic and acute exposure to Hg^0 can lead to a variety of long term or instantaneous health issues, a discussion after mapping allowed trainees to revisit the relationship between Hg^0 concentrations and health effects (Table 2).

Figure 2. Map generated of elemental mercury (Hg^0) concentrations during phase one of training in Georgetown, Guyana.

Table 2. Benchmark [Hg^0] in air and corresponding limits.

[Hg^0] (ng/m^3)	Agency	Description/Potential Health Effects	Reference
200	ATSDR [a]	Minimum risk level (MRL)	[73]
1000	ATSDR [a]	Recommended action level, residential setting	[74]
2000	Government of Peru	Air quality standards, Total gaseous mercury (TGM), not to exceed value over 24 h.	[69,75]
10,000	ATSDR [a]	Isolation of residential setting (evacuation, restricted access, etc.)	[74]
20,000	WHO [b]	Chronic exposure greater than or equal to this value can result in damage to the central nervous system	[30]
25,000	ACGIH [c]	Threshold limit value (TLV)	[76]
50,000	NIOSH [d]	Recommended exposure limit (REL), 10 h time-weighted average	[76]
100,000	NIOSH [d]/OSHA [e]	Acceptable ceiling concentration	[76]
670,000	USEPA [f]	Acute exposure guideline limit (AEGL) 2; 1 h, irreversible, serious, and/or long-lasting health effects may occur	[77]
2,200,000	USEPA [f]	Acute exposure guideline limit (AEGL) 3; 4 h, life-threatening effects or death	[77]
3,100,000	USEPA [f]	Acute exposure guideline limit (AEGL) 2; 10 min, irreversible, serious, and/or long-lasting health effects may occur	[77]
10,000,000	NIOSH [d]	Immediately dangerous to life or health (IDLH)	[78]

[a] Agency for Toxic Substances and Disease Registry; [b] World Health Organization; [c] American Conference of Governmental Industrial Hygienists; [d] National Institute for Occupational Safety and Health; [e] Occupational Safety and Health Administration; [f] United States Environmental Protection Agency.

Oftentimes, gold shops serve as a residence for the owner and his/her family, and as such environmental, residential, and industrial standards were highlighted. Reviewing these benchmark concentrations was particularly important when considering the fact that Guyana currently has no Hg emissions standards (Hg^0 or TGM); therefore, the data collected by the participants were put into context using international standards.

Due to a lack of ventilation, concentrations in gold shops during burning often exceeds these values, and exposure to Hg^0 vapor during burning can contribute to severe lung damage or death [31,47–49,79–81]. Two published cases of lung damage occurring during the processing of amalgams, one specific to a Guyanese miner, were discussed [48,49]. Trainees were reminded that the maps generated during this process were a snapshot in time; the data collected served as a record that Hg^0 was being emitted at these locations, but Hg^0 emissions from these sites could not be determined or even estimated using these protocols. As a screening protocol, this method is both rapid and effective.

To contrast the difference between Hg^0 concentrations emitted from Georgetown gold shops and the local GGB smelting facility, air concentrations surrounding the facility were monitored with a Lumex, and the VM-3000 was set up inside GGB and monitored for ≈4 h during burning operations. Concentrations on the street outside of the facility never exceeded 200 ng/m^3, and Hg^0 concentrations never exceeded 2800 ng/m^3 during smelting. A discussion was held regarding the value of screening a large area using the mapping protocol vs. fixed-point measurements at a gold shop. Trainees correctly concluded that both were useful techniques, but the data collected had very different uses and applications. For example, while mapping allows for the screening of an entire neighborhood or community, it only reflects the concentrations of Hg^0 over short time periods and only when the spectrometer is present. Fixed-position monitoring provides a time-weighted average over a longer time period, and thus is more reflective of Hg^0 concentrations in a given environment. However, individual fixed-position systems do not provide an overview of concentrations throughout a neighborhood or community.

3.3. Phase II: Training in Bartica

Training sessions in Bartica were built upon course material discussed during the Georgetown session, with the understanding that participants would be required to use all instrumentation in a real-world setting during the "do" phase. Training in Bartica occurred from 20–30 May, with a cohort of 18 persons (16 GGMC and 2 EPA) being trained throughout. Four additional persons representing NGOs and miners' unions attended training on select days. Although there was a three-day break from formal training for the observance of Guyanese Independence Day, trainers and trainees collected data throughout the entire training period.

Prior to departure from Georgetown, Mercer University and the GGMC agreed that successful completion of training would require (1) collection of data with the Lumex, VM-3000, and MTIP; (2) organization of data collected using the GPS-linked Lumex using a spreadsheet program; (3) generation of a map plotting [Hg^0] vs location using QGIS and/or Google Heat Maps; and (4) interpretation of the maps and [Hg^0] collected in the field using the MTIP. Progress in these areas were assessed by both Mercer University collaborators and select GGMC employees. Training would also occur on the collection of data in gold shops using the stationary VM-3000. However, many of the gold shops were small, and it was difficult for more than one person to be in a gold shop at a time. To this end, data were collected but training in fixed-position monitoring was not mandated for all participants. Similarly, using the XRF was not required for trainees, and therefore data collected during the training were not presented in this report.

A brief but general overview of the lectures in Guyana was conducted on the first day. Because Hg^0 vapor is both highly toxic and invisible, safety training specific to field work was conducted first. Trainees and trainers operating the MTIP were supplied with half-face masks with the appropriate filter for Hg^0 vapor, and taught how to fit and use the masks properly. The MTIP has a long wand that allows for early detection of elevated concentrations of Hg^0 vapor; the MTIP protects both the operator and the more sensitive Lumex, which is prone to memory effects at higher concentrations. In spite of the potential danger of Hg^0, perhaps the most dangerous component of monitoring Hg^0 emissions is traffic. Operators become so engrossed in what they are doing (or so distracted by conversation) they can accidentally walk into traffic. To this end, each portable spectrometer was operated by a team of three people. In the case of the Lumex, one carries the instrument and is responsible for directing the intake wand; another carries the computer and monitors data collection; and a third watches out for obstacles and safety hazards, ensures the connection is maintained between the Lumex and the computer, and ensures that an adequate pace is maintained throughout the monitoring period.

After safety training, pacing for walking with the instruments was practiced. The Lumex with a GPS unit records data in 1-s intervals. Walking too quickly decreases the number of readings per GPS coordinate and may disrupt the flow of air entering the Lumex. Both moving the Lumex and swinging the intake hose too quickly effectively alter the path length of the instrument. In our experience, this results in the recording of artificially low concentrations in contaminated areas near gold shops, although the effect is not noticeable in areas where only low, natural background levels of Hg^0 are found. Teams were also instructed to monitor the weather. As training was occurring during the rainy season, teams were instructed to preemptively save their data, shut down the instruments, and place them in plastic bags to be immediately brought back to their storage location.

After the initial training was completed, the 18 trainees and the trainers were divided into 4 teams distributed amongst the 3 portable spectrometers and the XRF. By the end of the training, five teams were active, with one team resting while the others worked. The majority of data collection occurred between 8:00 a.m. and 4:00 p.m. Each day, the data collected were downloaded to a computer, organized, and mapped using QGIS. As with data collection, these processes were initially demonstrated by the trainers but ultimately conducted by trainees.

3.3.1. Mapping of Central Bartica

Although maps were generated each day, the data collection process was slow, and even with two Lumex instruments it was impossible to map all of Bartica in one day. In addition, because of the number of trainees and the nature of the mapping protocol, it was necessary to walk the same paths through the town repeatedly. Although maps were generated at the end of each day, on the last day of the training period, all location vs. data collected were combined into a single spreadsheet. The maximum values of Hg^0 concentrations that fell on the Lumex calibration curve at each unique location were mapped (Figure 3).

The map clearly identified gold shops emitting Hg^0. Unlike previous work conducted in Peru and Ecuador, the concentrations determined on the street using the Lumex were significantly lower, with the exception of one occasion wherein the MTIP never exceeded 100,000 ng/m^3. This is attributed to the fact that many of the gold shops were considerably set back from the street and sidewalks, were often vented with high chimneys, and were not operating as frequently as in other ASGM communities. Numerous gold shop owners stated that business had been slowed by heavy rains and flooding in the interior that limited transportation to Bartica, while a few commented that the number of gold shops significantly decreased the amount of gold purchased at any one shop.

Figure 3. Map of Central Bartica, consisting of maximum Hg^0 concentrations found during Bartica training, May 2019. The majority of active gold shops were located within the red boxes.

3.3.2. Monitoring Hg^0 Concentrations During Burning/Smelting Operations at Gold Shops

Concentrations of Hg^0 inside two gold shops and at the Bartica branch of the GGB were recorded during the smelting of sponge gold using the VM-3000. Data collection at both gold shops was only allowed under the condition of anonymity. Due to the small size of the rooms in which burning occurred and the fact that there was no set timetable to burn gold, we made efforts to ensure that the monitoring did not compromise the safety of the instrument operators or gold shop employees. To this end, the VM-3000 was set to measure Hg^0 concentrations every second, and the data were stored to the onboard computer. As there is a finite amount of storage space on the VM-3000, at certain intervals, we downloaded data to a laptop computer. At all locations, the unit was placed on the floor of the room, 1–2 m from the opening of the fume hood where gold was smelted. Ideally, the spectrometer intake would be located at approximately face level to determine Hg^0 concentrations in locations where employees would be breathing; however, the unit would have been a tripping hazard in the confined spaces of the rooms.

The GGB facility in Bartica was evaluated the day before monitoring, and the facility office area was found to have Hg^0 of less than 1000 ng/m^3 as measured by the MTIP prior to burning. The burning room itself, which had been used ≈40 min prior, still had the ventilation system running and had a concentration of < 2000 ng/m^3. The following day, the VM-3000 was set up, and data recording commenced. The results of the study are found in Figure 4. The maximum concentration at floor level was determined to be 98,700 ng/m^3 during the study period, in which sponge gold was smelted once. Over the 7 h and 24 min of monitoring, the average Hg^0 concentration was 9590 ng/m^3. These Hg^0

concentrations were significantly higher than the Hg concentrations measured at the GGB facility in Georgetown, which had recently installed a modern ventilation and capture system. Guyana currently has no laws related to ceiling indoor Hg^0 concentrations in industrial environs, nor a time-weighted average for employee exposure. This was an initial assessment of $[Hg^0]$ in this GGB facility. As only one smelting episode was recorded on one day, it is impossible to draw long-term conclusions from the data presented here. It is possible that the Occupational Safety and Health Administration permissible exposure limit (OSHA PEL) of 100,000 ng/m^3 is exceeded during burning at the face of the fume hood. The same holds true for the National Institute for Occupational Safety and Health recommended exposure limit (NIOSH REL) of 50,000 ng/m^3 (TWA) over an 8-h workday. However, employees in the room were equipped with appropriate personal protective equipment including laboratory coats and half-face masks, and on the basis of the collected data it is clear that the GGB sites evaluated during this study had significantly less Hg^0 contamination than the gold shops assessed. Future monitoring should be designed to monitor Hg^0 at approximate human face level. In addition, while no Hg^0 measured outside was directly attributed to the GGB Bartica facility, future assessment should include monitoring of the effluent emitted directly from the facility's ventilation system to determine the concentrations of Hg^0 emitted during burning.

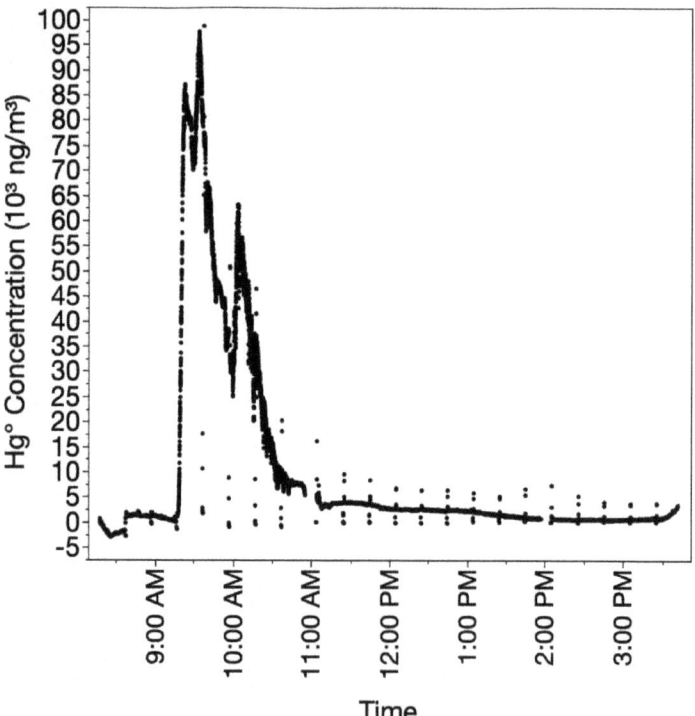

Figure 4. Hg^0 concentrations recorded during the smelting of sponge gold in the Guyana Gold Board (GGB) processing facility in Bartica, Guyana.

A gold shop had been identified during the prior day's measurements as having significantly elevated concentrations of Hg^0 at the front opening. The owners openly allowed the shop and connected waiting area to be assessed for Hg^0 concentrations. They stated that their ventilation system was installed by a private contractor and cost $2500 (USD). Concentrations in the waiting room exceeded 50,000 ng/m^3, as measured by the MTIP, and the burn room was highly fluxional, with concentrations ranging between 100,000 ng/m³ and 200,000 ng/m³. The owners reported that business had been slow, and they had not burned in three days.

The following morning, the owners contacted us and stated that they would be receiving sponge gold and would be smelting the next day. They granted permission to set up the VM-3000, but stated that we could not be at the location during smelting. As such, it was impossible to link the activities in the gold shop to Hg^0 concentrations, but it was ascertained that multiple pieces of sponge gold were reheated, and the owners combined and smelted the gold during the observation period. The results can be found in Figure 5.

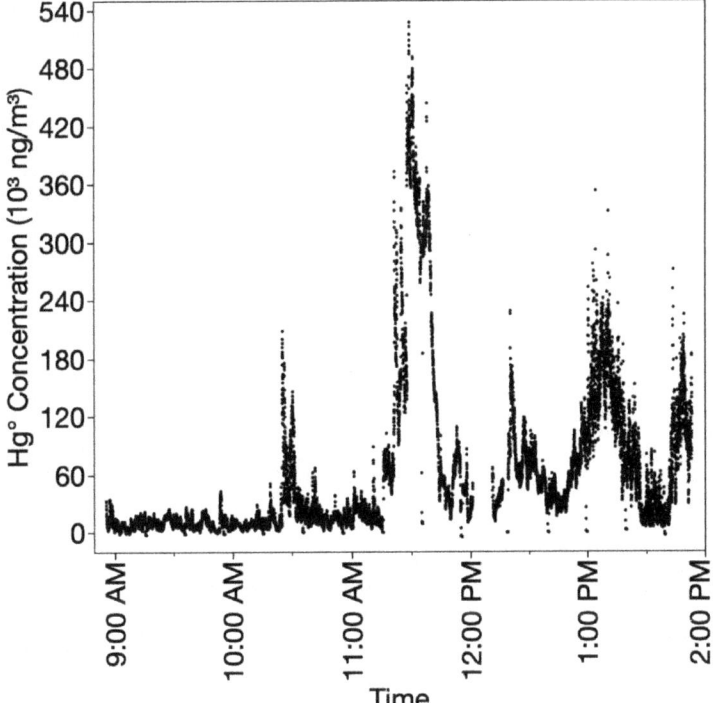

Figure 5. Hg^0 concentrations recorded at a gold shop in Bartica, Guyana.

The maximum concentration of Hg^0 as measured during burning was 527,500 ng/m³, with an average concentration of 58,400 ng/m³ over the ≈5-h monitoring period. It is clear that concentrations of Hg^0 in this gold shop exceeded safe levels; however, it should be noted that this gold shop had a ventilation system that passed air from the fume hood through a 55-gallon drum filled with water, and the vast majority of Hg^0 was removed from the room. The owners claimed that the system captured Hg^0 but had never recovered it. Concentrations during burning at the exhaust exceeded 2,000,000 ng/m³, the detection limit of the MTIP. Other gold shops had a variety of these ventilation/capture systems, but all seem to be custom-made and unique. This may lead to future issues in monitoring the effectiveness of these systems should an emissions standard be developed.

Because the mapping process was highly visible, with dozens of trainees and trainers walking on the streets with unwieldy instrumentation, we were approached by numerous citizens, miners, and gold buyers who were interested in what we were doing. One gold shop owner reported that their neighbors were complaining that their shop was venting into the courtyard behind the building into an area where a young child was sleeping. As a result, the gold shop was assessed and Hg^0 concentrations were determined during the smelting of sponge gold into a single ingot. The second gold shop was located on the ground floor of a two-story building that also served as a residence and a business unrelated to mining. Behind the building was a courtyard and another residence. Earlier in the day, the owner placed a 90° angle bend at the top of the exhaust pipe, which extended above the roof of the building. The intent was to direct the vented gas away from the courtyard and toward the street. The monitoring of the smelting can be found in Figure 6. Similar to the previous informal gold shop, concentrations over the brief monitoring exceeded the OSHA PEL with a maximum concentration of 302,200 ng/m^3.

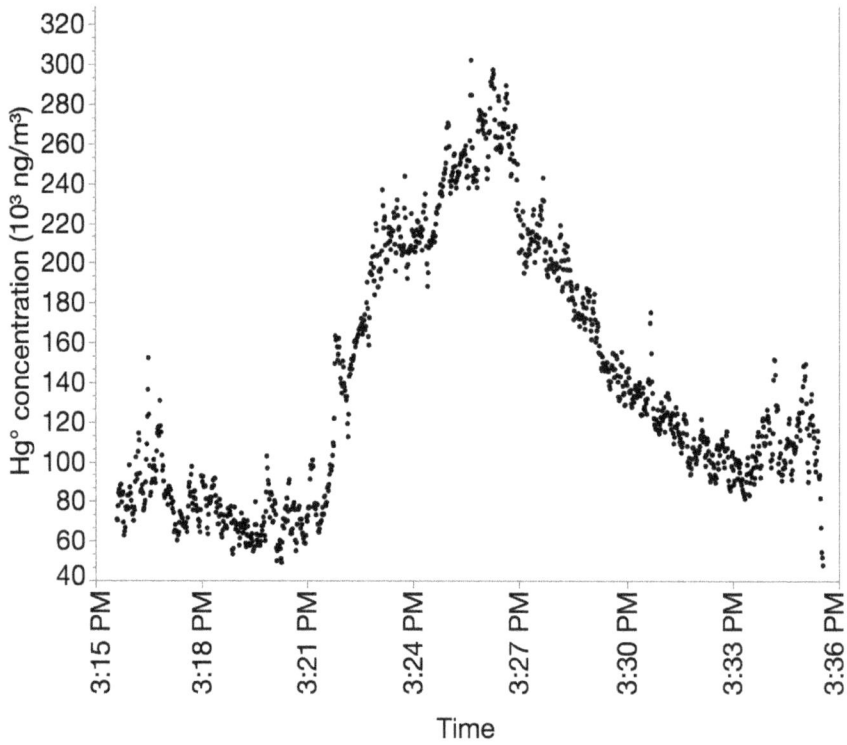

Figure 6. Hg^0 concentrations recorded at a gold shop in Bartica, Guyana.

The gold shop contained a single fume hood that vented through the wall behind it. A pipe extended through the wall and was directed toward the ground and into a box containing a fan. The two-story exhaust pipe extended up from this box. During smelting, concentrations rapidly exceeded 650,000 ng/m^3 within 30 cm from the pipe/box junctures; the MTIP was removed from this area immediately to prevent contamination. However, 1 m away, concentrations were less than 25,000 ng/m^3, and were less than 5000 ng/m^3 within 3 m at the entrance of the courtyard. Meanwhile, an unsuspecting Lumex team who had stopped mapping for the day and was waiting on the opposite side of the street ≈25 m from the front of the building recorded a rapid increase to >25,000 ng/m^3 and immediately left the area. During monitoring, it seemed as though the 90° bend on the exhaust pipe served to redirect the Hg^0 vapor onto the street and away from the building. It is important to note that this system may not effectively move the Hg^0 away from the residence, and the vapor may have been carried away from the gold shop during the observation time by the prevailing wind from the river.

3.3.3. The Relationship Between Gold Shops and Hg^0 Contamination in Bartica

Not surprisingly, all Hg^0 vapor contamination in Bartica was found to originate at gold shops during phase 2 of training, and all elevated concentrations can be traced back to either (1) Hg^0 contamination from previous smelting activities in gold shops, (2) the burning of amalgams, (3) the reheating of processed amalgams, or (4) smelting of gold. There are seven GGB licensed gold dealers in Bartica [82] and over 30 gold shops, although some were not actively burning during the training period. In spite of the large number of gold shops, Hg^0 concentrations in public areas such as streets and sidewalks were considerably lower when compared to previous work mapping conducted in Peru.

There are numerous reasons that the maps generated in Bartica display relatively less Hg^0 than in other communities. As previously mentioned, less burning occurred during the rainy season due to flooding. As a result, it was less likely to find that gold shops were open throughout the entire day. During training many gold shops were closed from ≈2:00–4:00 p.m., as were many other stores. In addition, many stores closed when rain came. Gold shop owners that chose to close their shops may reduce the amount of Hg^0 emanating from the shop, thus decreasing the likelihood that Hg^0 would be detected if a Lumex team passed. Many Guyanese gold shops were also set back further from the street and sidewalk than Peruvian shops, and most gold shops had burning stations located in the back of the business away from the street. Unlike gold shops or burning facilities in Peru, there were no gold shops in Bartica that vented directly onto the street through horizontal piping. Many shops in Guyana had vertical exhaust stacks that may lead to greater dilution of Hg^0 vapor than those in other ASGM communities. However, as highlighted at the second gold shop, the effect of tall chimneys on Hg^0 emissions from gold shops may lead to wider distribution of Hg^0 pollution and affect mapping and identification activities. Finally, certain gold shops in Guyana had conditioned air that led to exterior windows and doors being closed, whereas in Peru, the front of most shops were open. The closing of shops to maintain a cool temperature may lead to higher concentrations of Hg^0 trapped inside the burning area. The effect of tall chimneys and air conditioning should be investigated at a later date.

Previous work in Ecuador and Peru had demonstrated that although Hg^0 concentrations may vary from day to day, that mapping could always be used to determine the locations of gold shops. It became apparent during training that this was not to be the case in Bartica; there were mercury emissions not noted at a given location one day, and yet mercury emissions appeared on another day of training (Figure 7). Therefore, the mapping technique presented to trainees is noted as an effective screening tool for identifying gold shops that is limited because it only records concentrations at a given time. For this reason, trainees were routinely told that mapping represented a "snapshot in time". Trainees were reminded that mapping cannot be carried out over a single day, but must be comprised of data collected over multiple days.

3.4. Phase III: Analysis and Dissemination of Results

During joint planning sessions between Mercer and the GGMC, a GGMC representative stated that although employees were capable of collecting data in the field, a potential area for growth could be interpretation of the data collected and the generation of the final report. While all 18 full-time trainees (1) were instructed in Guyana and Bartica; (2) participated in the collection of data using the Lumex, MTIP, and XRF; (3) mapped at least one day's worth of collected data using Microsoft Excel and quantum geographic information system (QGIS); (4) interpreted the map and referenced it with their observations in the field; and (5) assisted or were presented with the final findings of the mapping work, it was logistically infeasible to engage all trainees and trainers in the production and review of the final report. As a result of this conversation, selected GGMC employees participated in the final interpretation of data and writing of the report. These trainees were selected because of their dedication to the collection and interpretation of data. These trainees were ultimately able to generate maps independent of trainers, and as such, were identified as potentially being able to lead future internal training for the GGMC. On the final day of training, trainees assisted in the generation of the total map and participated in a discussion related to the findings of the program.

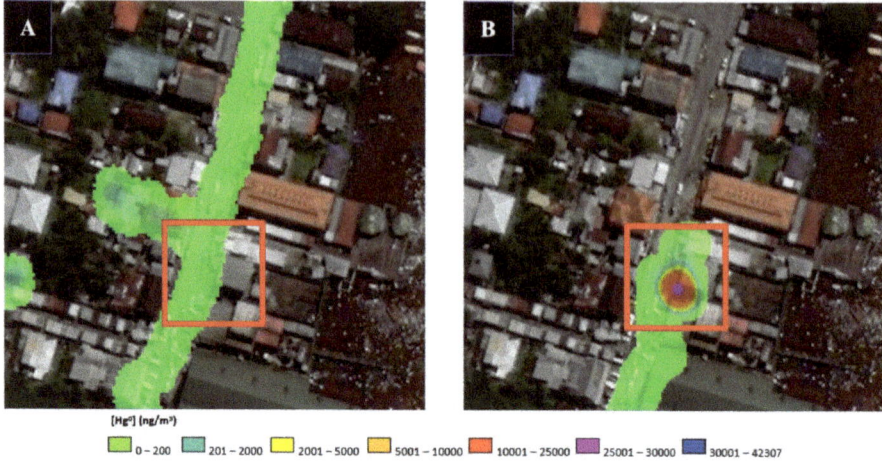

Figure 7. Panels (**A**,**B**) are daily maps generated by trainees. No burning activity (panel (**A**)) may not be reflective of gold shop activity. When sponge gold or amalgams are heated in the same location in subsequent days, Hg^0 is clearly visible on the map (panel (**B**)).

Upon the generation of the drafted report, the trainees were asked to bridge the scientific findings with their experiences during the training. In addition, they were asked to place the technical and scientific findings in the context of Guyanese mining practices.

4. Conclusions and Future Directions

A three-phase training program was established from a collaboration between a U.S. university and the GGMC. The first phase of the training encompassed the use of Hg^0 in ASGM in the context of science, the environment, and policy, in particular the Minamata Convention. Participants were introduced to a suite of analytical instrumentation to assess Hg^0 contamination originating from gold shops in the capital city of Georgetown. During the second phase of training in Bartica, data were collected and mapped that quantified Hg^0 pollution at gold shops. In addition, concentrations of Hg^0 were measured inside the Guyana Gold Board facilities. During monitoring periods, GGB facilities remained under the OSHA PEL guideline of 100,000 ng/m^3, but the gold shops produced Hg^0 concentrations that exceeded the PEL. Furthermore, these gold shops were found to vent concentrations of Hg^0 exceeding 1,000,000 ng/m^3, with one shop exceeding 2,000,000 ng/m^3, the detection limit of the MTIP. During the third and final phase of the project, select representatives from the GGMC were asked to assist in the preparation of a final report, reinforcing learned concepts and requiring further interpretation of the results.

Because the GGMC has identified the need for the development of standards for the emission of Hg^0, the data collected during this training may serve as a baseline for deciding emissions limits and may assist in the generation of national air quality standards. A delicate balance must be reached between profits from gold mining and protecting human and environmental health. The reality is that existing laws in other countries may not apply to areas of high contamination such as gold shops. This program also identifies sources of Hg^0 to the atmosphere, but it does not provide insight into the ultimate fate of this Hg^0. The maps generated indicate that Hg^0 is rapidly diluted once it leaves the gold shop. The reality is that this Hg^0 must go somewhere, entering the atmosphere and/or depositing on the ground or surface of buildings, and thus would be inaccessible to the sampling probes of the spectrometers. Whether Hg^0 from these gold shops is deposited locally, regionally, or globally has yet to be determined. In other areas, inexpensive passive air samplers have been used to quantify Hg^0 in the air over extended periods of time [83–86]. Future mapping that incorporates data from passive air samplers may add to our understanding of the fate of Hg^0 emitted from gold shops. This is a global challenge, and as such, will be faced not only by Guyana but by all signatory nations of the Minamata Convention with ASGM activities.

Author Contributions: Conceptualization: A.M.K.; investigation: S.T.B., L.L.B., S.S.A., S.T.T., T.E.G., G.K.M., A.V.E., K.M.H., D.C.L., and A.M.K.; methodology: S.T.B., K.M.H., D.C.L., C.S.S., and A.M.K.; project administration: L.L.B. and A.M.K.; resources: A.M.K.; supervision: L.L.B., S.S.A., and A.M.K.; visualization: S.T.B., G.K.M., A.V.E., and A.M.K.; writing—original draft: S.T.B., L.L.B., S.S.A., S.T.T., T.E.G., G.K.M., A.V.E., K.M.H., C.S.S., and A.M.K.; writing—review and editing: S.T.B., L.L.B., S.S.A., S.T.T., T.E.G., D.C.L., C.S.S., and A.M.K. All authors have read and agreed to the published version of the manuscript.

Funding: This research received no external funding.

Acknowledgments: The authors acknowledge Guyana's Ministry of Natural Resources, the Guyana Geology and Mines Commission, and the Mercer on Mission program for financial and logistical support for this project. Madison Ayers, Hailey Christian, Alaina Dawson, Owen House, Kayla Kelley, Alina McCue, Kyle Powell, and Sutton Scarboro are acknowledged for assisting with in-country data collection and training.

Conflicts of Interest: The authors declare no conflict of interest.

References

1. Seccatore, J.; Veiga, M.; Origliasso, C.; Marin, T.; De Tomi, G. An Estimation of the Artisanal Small-Scale Production of Gold in the World. *Sci. Total Environ.* **2014**, *496*, 662–667. [CrossRef] [PubMed]
2. Cortés-McPherson, D. Expansion of Small-Scale Gold Mining in Madre de Dios: 'Capital Interests' and the Emergence of a New Elite of Entrepreneurs in the Peruvian Amazon. *Extr. Ind. Soc.* **2019**, *6*, 382–389. [CrossRef]
3. Hentschel, T.; Hruschka, F.; Priester, M. *Artisanal and Small-Scale Mining: Challenges and Opportunities*; International Institute for Environment and Development: London, UK, 2004.
4. Hilson, G.; Zolnikov, T.R.; Ortiz, D.R.; Kumah, C. Formalizing Artisanal Gold Mining under the Minamata Convention: Previewing the Challenge in Sub-Saharan Africa. *Environ. Sci. Policy* **2018**, *85*, 123–131. [CrossRef]
5. Hilson, G.; Pardie, S. Mercury: An Agent of Poverty in Ghana's Small-Scale Gold-Mining Sector? *Resour. Policy* **2006**, *31*, 106–116. [CrossRef]
6. Hilson, G.M. (Ed.) *The Socio-Economic Impacts of Artisanal and Small-Scale Mining in Developing Countries*; Taylor & Francis: Oxfordshire, UK, 2003.
7. Hruschka, F.; Echavarria, C. *Rock Solid Chances for Responsible Artisanal Mining*; Series on Responsible ASM; Alliance for Responsible Mining: Medellin, Columbia, 2011.
8. Jønsson, J.B.; Charles, E.; Kalvig, P. Toxic Mercury versus Appropriate Technology: Artisanal Gold Miners' Retort Aversion. *Resour. Policy* **2013**, *38*, 60–67. [CrossRef]
9. Quiroga, E.R. The Case of Artisanal Mining in Bolivia: Local Participatory Development and Mining Investment Opportunities. *Nat. Resour. Forum* **2002**, *26*, 127–139. [CrossRef]
10. Spiegel, S.J. Socioeconomic Dimensions of Mercury Pollution Abatement: Engaging Artisanal Mining Communities in Sub-Saharan Africa. *Ecol. Econ.* **2009**, *68*, 3072–3083. [CrossRef]
11. Caballero Espejo, J.; Messinger, M.; Román-Dañobeytia, F.; Ascorra, C.; Fernandez, L.E.; Silman, M. Deforestation and Forest Degradation Due to Gold Mining in the Peruvian Amazon: A 34-Year Perspective. *Remote Sens.* **2018**, *10*, 1903. [CrossRef]
12. Dezécache, C.; Faure, E.; Gond, V.; Salles, J.-M.; Vieilledent, G.; Hérault, B. Gold-Rush in a Forested El Dorado: Deforestation Leakages and the Need for Regional Cooperation. *Environ. Res. Lett.* **2017**, *12*, 034013. [CrossRef]
13. Deikumah, J.P.; McAlpine, C.A.; Maron, M. Mining Matrix Effects on West African Rainforest Birds. *Biol. Conserv.* **2014**, *169*, 334–343. [CrossRef]
14. Estrada, A.; Garber, P.A.; Chaudhary, A. Expanding Global Commodities Trade and Consumption Place the World's Primates at Risk of Extinction. *PeerJ* **2019**, *7*, e7068. [CrossRef]
15. Cummings, A.R.; Piquero, A.R.; Accetturo, P.W. Insights into the Nature and Spatial Distribution of Guyanese Crime. *Deviant Behav.* **2018**, *39*, 20–34. [CrossRef]

16. Hilson, G.; Laing, T. Gold Mining, Indigenous Land Claims and Conflict in Guyana's Hinterland. *J. Rural Stud.* **2017**, *50*, 172–187. [CrossRef]
17. Calys-Tagoe, B.N.L.; Ovadje, L.; Clarke, E.; Basu, N.; Robins, T. Injury Profiles Associated with Artisanal and Small-Scale Gold Mining in Tarkwa, Ghana. *Int. J. Environ. Res. Public Health* **2015**, *12*, 7922–7937. [CrossRef]
18. Calys-Tagoe, B.N.L.; Clarke, E.; Robins, T.; Basu, N. A Comparison of Licensed and Un-Licensed Artisanal and Small-Scale Gold Miners (ASGM) in Terms of Socio-Demographics, Work Profiles, and Injury Rates. *BMC Public Health* **2017**, *17*, 862. [CrossRef]
19. Esdaile, L.J.; Chalker, J.M. The Mercury Problem in Artisanal and Small-Scale Gold Mining. *Chem. Eur. J.* **2018**, *24*, 6905–6916. [CrossRef]
20. Gibb, H.; O'Leary, K.G. Mercury Exposure and Health Impacts among Individuals in the Artisanal and Small-Scale Gold Mining Community: A Comprehensive Review. *Environ. Health Perspect.* **2014**, *122*, 667–672. [CrossRef]
21. Hylander, L.D. *Gold and Amalgams: Environmental Pollution and Health Effects*; Elsevier: Burlington, NJ, USA, 2011.
22. Markham, K.E.; Sangermano, F. Evaluating Wildlife Vulnerability to Mercury Pollution from Artisanal and Small-Scale Gold Mining in Madre de Dios, Peru. *Trop. Conserv. Sci.* **2018**, *11*. [CrossRef]
23. Martinez, G.; McCord, S.A.; Driscoll, C.T.; Todorova, S.; Wu, S.; Araújo, J.F.; Vega, C.M.; Fernandez, L.E. Mercury Contamination in Riverine Sediments and Fish Associated with Artisanal and Small-Scale Gold Mining in Madre de Dios, Peru. *Int. J. Environ. Res. Public Health* **2018**, *15*, 1584. [CrossRef]
24. Moreno-Brush, M.; McLagan, D.S.; Biester, H. Fate of Mercury from Artisanal and Small-Scale Gold Mining in Tropical Rivers: Hydrological and Biogeochemical Controls. A Critical Review. *Crit. Rev. Environ. Sci. Technol.* **2020**, *50*, 437–475. [CrossRef]
25. Steckling, N.; Tobollik, M.; Plass, D.; Hornberg, C.; Ericson, B.; Fuller, R.; Bose-O'Reilly, S. Global Burden of Disease of Mercury Used in Artisanal Small-Scale Gold Mining. *Ann. Glob. Health* **2017**, *83*, 234–247. [CrossRef] [PubMed]
26. Basu, N.; Horvat, M.; Evers, D.C.; Zastenskaya, I.; Weihe, P.; Tempowski, J. A State-of-the-Science Review of Mercury Biomarkers in Human Populations Worldwide between 2000 and 2018. *Environ. Health Perspect.* **2018**, *126*, 106001. [CrossRef]
27. Bose-O'Reilly, S.; Lettmeier, B.; Gothe, R.M.; Beinhoff, C.; Siebert, U.; Drasch, G. Mercury as a Serious Health Hazard for Children in Gold Mining Areas. *Environ. Res.* **2008**, *107*, 89–97. [CrossRef] [PubMed]
28. United Nations Environmental Program. *Global Mercury Assessment 2013: Sources, Emissions, Releases and Environmental Transport*; United Nations Pubns: Geneva, Switzerland, 2013.
29. Environment, U.N. Global Mercury Assessment. 2018. Available online: http://www.unenvironment.org/resources/publication/global-mercury-assessment-2018 (accessed on 29 October 2019).
30. Risher, J.; World Health Organization; United Nations Environment Programme; International Labour Organisation; Inter-Organization Programme for the Sound Management of Chemicals; International Program on Chemical Safety. *Elemental Mercury and Inorganic Mercury Compounds: Human Health Aspects*; World Health Organization: Geneva, Switzerland, 2003.
31. Eisler, R. Health Risks of Gold Miners: A Synoptic Review. *Environ. Geochem. Health* **2003**, *25*, 325–345. [CrossRef] [PubMed]
32. Fernandes Azevedo, B.; Barros Furieri, L.; Peçanha, F.M.; Wiggers, G.A.; Frizera Vassallo, P.; Ronacher Simões, M.; Fiorim, J.; Rossi de Batista, P.; Fioresi, M.; Rossoni, L.; et al. Toxic Effects of Mercury on the Cardiovascular and Central Nervous Systems. Available online: https://www.hindawi.com/journals/bmri/2012/949048/citations/ (accessed on 14 April 2019). [CrossRef]
33. Clarkson, T.W.; Magos, L. The Toxicology of Mercury and Its Chemical Compounds. *Crit. Rev. Toxicol.* **2006**, *36*, 609–662. [CrossRef]
34. Obrist, D.; Kirk, J.L.; Zhang, L.; Sunderland, E.M.; Jiskra, M.; Selin, N.E. A Review of Global Environmental Mercury Processes in Response to Human and Natural Perturbations: Changes of Emissions, Climate, and Land Use. *Ambio* **2018**, *47*, 116–140. [CrossRef]
35. Driscoll, C.T.; Mason, R.P.; Chan, H.M.; Jacob, D.J.; Pirrone, N. Mercury as a Global Pollutant: Sources, Pathways, and Effects. *Environ. Sci. Technol.* **2013**, *47*, 4967–4983. [CrossRef]
36. Clifford, M.J. Pork Knocking in the Land of Many Waters: Artisanal and Small-Scale Mining (ASM) in Guyana. *Resour. Policy* **2011**, *36*, 354–362. [CrossRef]

37. Chabrol, D. Gold Board Wants to Buy "All Gold" Directly from Miners. Available online: https://demerarawaves.com/2019/02/25/gold-board-wants-to-buy-all-gold-directly-from-miners/ (accessed on 6 July 2020).
38. Laws of Guyana. Chapter 66:01 Guyana Gold Board Act. Available online: http://www.guyaneselawyer.com/lawsofguyana/Laws/cap6601.pdf (accessed on 6 July 2020).
39. Pasha, S.; Wenner, M.D.; Clarke, D. Toward the Greening of the Gold Mining Sector of Guyana: Transition Issues and Challenges; Technical Note IDB-TN-1290; Inter-American Development Bank. Available online: https://publications.iadb.org/publications/english/document/Toward-the-Greening-of-the-Gold-Mining-Sector-of-Guyana-Transition-Issues-and-Challenges.pdf (accessed on 6 July 2020).
40. Staff Editor. Around 15,000 Ozs Gold Smuggled Each Week—Trotman. Available online: https://www.stabroeknews.com/2016/01/06/news/guyana/around-15000-ozs-gold-smuggled-week-trotman/ (accessed on 6 July 2020).
41. Canterbury, D.C. Natural Resources Extraction and Politics in Guyana. *Extr. Ind. Soc.* **2016**, *3*, 690–702. [CrossRef]
42. Hilson, G.; Laing, T. Guyana Gold: A Unique Resource Curse? *J. Dev. Stud.* **2017**, *53*, 229–248. [CrossRef]
43. Cordy, P.; Veiga, M.; Crawford, B.; Garcia, O.; Gonzalez, V.; Moraga, D.; Roeser, M.; Wip, D. Characterization, Mapping, and Mitigation of Mercury Vapour Emissions from Artisanal Mining Gold Shops. *Environ. Res.* **2013**, *125*, 82–91. [CrossRef] [PubMed]
44. Santa Rosa, R.M.; Müller, R.C.; Alves, C.N.; Sarkis, J.E.D.S.; Bentes, M.H.D.S.; Brabo, E.; de Oliveira, E.S. Determination of Total Mercury in Workers' Urine in Gold Shops of Itaituba, Pará State, Brazil. *Sci. Total Environ.* **2000**, *261*, 169–176. [CrossRef]
45. Wip, D.; Warneke, T.; Petersen, A.K.; Notholt, J.; Temme, C.; Kock, H.; Cordy, P. Urban Mercury Pollution in the City of Paramaribo, Suriname. *Air Qual. Atmos. Health* **2011**, *6*, 205–213. [CrossRef]
46. Cordy, P.; Veiga, M.M.; Salih, I.; Al-Saadi, S.; Console, S.; Garcia, O.; Mesa, L.A.; Velásquez-López, P.C.; Roeser, M. Mercury Contamination from Artisanal Gold Mining in Antioquia, Colombia: The World's Highest per Capita Mercury Pollution. *Sci. Total Environ.* **2011**, *410–411*, 154–160. [CrossRef]
47. De Lacerda, L.D.; Salomons, W. *Mercury from Gold and Silver Mining: A Chemical Time Bomb?* Springer: Berlin/Heidelberg, Germany, 1998.
48. Lilis, R.; Miller, A.; Lerman, Y. Acute Mercury Poisoning with Severe Chronic Pulmonary Manifestations. *Chest* **1985**, *88*, 306–309. [CrossRef]
49. Levin, M.; Jacobs, J.; Polos, P.G. Acute Mercury Poisoning and Mercurial Pneumonitis from Gold Ore Purification. *Chest* **1988**, *94*, 554–556. [CrossRef]
50. United Nations Treaty Collection. Minamatata Convention on Mercury. Available online: https://treaties.un.org/Pages/ViewDetails.aspx?src=IND&mtdsg_no=XXVII-17&chapter=27&clang=_en (accessed on 6 July 2020).
51. United Nations Environmental Programme. Mercury Convention Texts and Annexes. Available online: http://www.mercuryconvention.org/Convention/Text/tabid/3426/language/en-US/Default.aspx (accessed on 22 April 2019).
52. Clifford, M.J. Future Strategies for Tackling Mercury Pollution in the Artisanal Gold Mining Sector: Making the Minamata Convention Work. *Futures* **2014**, *62*, 106–112. [CrossRef]
53. Eriksen, H.H.; Perrez, F.X. The Minamata Convention: A Comprehensive Response to a Global Problem. *Rev. Eur. Comp. Int. Environ. Law* **2014**, *23*, 195–210. [CrossRef]
54. Evers, D.C.; Keane, S.E.; Basu, N.; Buck, D. Evaluating the Effectiveness of the Minamata Convention on Mercury: Principles and Recommendations for next Steps. *Sci. Total Environ.* **2016**, *569–570*, 888–903. [CrossRef]
55. Gustin, M.S.; Evers, D.C.; Bank, M.S.; Hammerschmidt, C.R.; Pierce, A.; Basu, N.; Blum, J.; Bustamante, P.; Chen, C.; Driscoll, C.T.; et al. Importance of Integration and Implementation of Emerging and Future Mercury Research into the Minamata Convention. *Environ. Sci. Technol.* **2016**, *50*, 2767–2770. [CrossRef]
56. Selin, H.; Keane, S.E.; Wang, S.; Selin, N.E.; Davis, K.; Bally, D. Linking Science and Policy to Support the Implementation of the Minamata Convention on Mercury. *Ambio* **2018**, *47*, 198–215. [CrossRef] [PubMed]
57. Selin, H. Global Environmental Law and Treaty-Making on Hazardous Substances: The Minamata Convention and Mercury Abatement. *Glob. Environ. Politics* **2014**, *14*, 1–19. [CrossRef]

58. Zolnikov, T.R.; Ramirez Ortiz, D. A Systematic Review on the Management and Treatment of Mercury in Artisanal Gold Mining. *Sci. Total Environ.* **2018**, *633*, 816–824. [CrossRef] [PubMed]
59. Jenkins, R. X-ray Techniques: Overview. In *Encyclopedia of Analytical Chemistry*; American Cancer Society: Hoboken, NJ, USA, 2006. [CrossRef]
60. Rousseau, R.M. The Quest for a Fundamental Algorithm in X-Ray Fluorescence Analysis and Calibration. *Open Spectrosc. J.* **2009**, *3*. [CrossRef]
61. Rousseau, R.; Boivin, J.A. The Fundamental Algorithm: A Natural Extension of Sherman Equation. *Rigaku J.* **1998**, *15*, 13–28.
62. Van Sprang, H.A. Fundamental Parameter Methods in XRF Spectroscopy. *Adv. X-ray Anal.* **2000**, *42*, 1–10.
63. Thomsen, V. Basic Fundamental Parameters in X-Ray Fluorescence. *Spectroscopy* **2007**, *22*, 46–50.
64. Ohio Lumex Co., Inc. *RA-915M Mercury Analyzer Operation Manual B0100-00-00-00-00 OM.*; Ohio Lumex Co., Inc.: Twinsburg, OH, USA, 2011.
65. Moody, K.H.; Hasan, K.M.; Aljic, S.; Blakeman, V.M.; Hicks, L.P.; Loving, D.C.; Moore, M.E.; Hammett, B.S.; Silva-González, M.; Seney, C.S.; et al. Mercury Emissions from Peruvian Gold Shops: Potential Ramifications for Minamata Compliance in Artisanal and Small-Scale Gold Mining Communities. *Environ. Res.* **2020**, *182*, 109042. [CrossRef]
66. Drace, K.; Kiefer, A.M.; Veiga, M.M.; Williams, M.K.; Ascari, B.; Knapper, K.A.; Logan, K.M.; Breslin, V.M.; Skidmore, A.; Bolt, D.A.; et al. Mercury-Free, Small-Scale Artisanal Gold Mining in Mozambique: Utilization of Magnets to Isolate Gold at Clean Tech Mine. *J. Clean. Prod.* **2012**, *32*, 88–95. [CrossRef]
67. Tsang, V.W.L.; Lockhart, K.; Spiegel, S.J.; Yassi, A. Occupational Health Programs for Artisanal and Small-Scale Gold Mining: A Systematic Review for the WHO Global Plan of Action for Workers' Health. *Ann. Glob. Health* **2019**, *85*, 128. [CrossRef]
68. Appel, P.W.U.; Na-Oy, L.D. Mercury-Free Gold Extraction Using Borax for Small-Scale Gold Miners. *J. Environ. Prot.* **2014**, *5*, 493–499. [CrossRef]
69. Sistema Nacional de Informacíon Ambiental. Aprueban Estándares de Calidad Ambiental (ECA) para Aire y establecen Disposiciones Complementarias. Available online: https://sinia.minam.gob.pe/normas/aprueban-estandares-calidad-ambiental-eca-aire-establecen-disposiciones (accessed on 15 April 2019).
70. Norma Técnica Peruano 900.068 (NTP 900.068). *Monitoreo de Calidad Ambiental. Calidad del Aire. Método Normalizado Para la Determinación del Mercurio Gaseoso Total*; Dirección de Normalización—INACAL: Lima, Peru, 2016.
71. Kiefer, A.M.; Seney, C.S.; Boyd, E.A.; Smith, C.; Shivdat, D.S.; Matthews, E.; Hull, M.W.; Bridges, C.C.; Castleberry, A. Chemical Analysis of Hg0-Containing Hindu Religious Objects. *PLoS ONE* **2019**, *14*, e0226855. [CrossRef] [PubMed]
72. Adam, K. Parad Items Readily Emit Mercury Vapor. Available online: https://www.youtube.com/watch?v=vV3fSXpt0-Q (accessed on 14 April 2019).
73. Agency for Toxic Substances & Disease Registry. ATSDR-Toxicological Profile: Mercury. Available online: http://www.atsdr.cdc.gov/toxprofiles/tp.asp?id=115&tid=24 (accessed on 26 June 2015).
74. Agency for Toxic Substances and Disease Registry. *Action Levels for Elemental Mercury Spills: Chemical-Specific Health Consultation for Joint EPA/ATSDR National Mercury Cleanup Policy Workgroup*; Agency for Toxic Substances and Disease Registry: Atlanta, GA, USA, 2012.
75. Plataforma Digital Única del Estado Peruano. Decreto Supremo N° 10-2019-MINAM. Available online: https://www.gob.pe/institucion/minam/normas-legales/363557-10-2019-minam (accessed on 12 December 2019).
76. United States Department of Labor: Occupational Safety and Health Administration. OSHA Annotated PELs. Available online: https://www.osha.gov/dsg/annotated-pels/tablez-2.html (accessed on 26 June 2015).
77. US EPA. Mercury Vapor Results-AEGL Program. Available online: https://www.epa.gov/aegl/mercury-vapor-results-aegl-program (accessed on 7 April 2019).
78. Centers for Disease Control and Prevention: The National Institute for Occupational Safety and Health. CDC-Immediately Dangerous to Life or Health Concentrations (IDLH): Mercury Compounds [Except (Organo) Alkyls] (as Hg)-NIOSH Publications and Products. Available online: http://www.cdc.gov/niosh/idlh/7439976.html (accessed on 26 June 2015).
79. Hacon, S.; Rochedo, E.R.; Campos, R.; Rosales, G.; Lacerda, L.D. Risk Assessment of Mercury in Alta Floresta. Amazon Basin-Brazil. *Water Air Soil Pollut.* **1997**, *97*, 91–105. [CrossRef]

80. Drake, P.L.; Rojas, M.; Reh, C.M.; Mueller, C.A.; Jenkins, F.M. Occupational Exposure to Airborne Mercury during Gold Mining Operations near El Callao, Venezuela. *Int. Arch. Occup. Environ. Health* **2001**, *74*, 206–212. [CrossRef]
81. Donoghue, A.M. Mercury Toxicity Due to the Smelting of Placer Gold Recovered by Mercury Amalgam. *Occup. Med. (Lond.)* **1998**, *48*, 413–415. [CrossRef]
82. Guyana Gold and Diamond Miners Association. Authorized Gold Dealers of the Guyana Gold Board. Available online: https://ggdma.com/pressnews/authorized-gold-dealers-of-the-guyana-gold-board/ (accessed on 8 September 2019).
83. McLagan, D.S.; Mitchell, C.P.J.; Huang, H.; Lei, Y.D.; Cole, A.S.; Steffen, A.; Hung, H.; Wania, F. A High-Precision Passive Air Sampler for Gaseous Mercury. *Environ. Sci. Technol. Lett.* **2016**, *3*, 24–29. [CrossRef]
84. McLagan, D.S.; Mazur, M.E.E.; Mitchell, C.P.J.; Wania, F. Passive Air Sampling of Gaseous Elemental Mercury: A Critical Review. *Atmos. Chem. Phys.* **2016**, *16*, 3061–3076. [CrossRef]
85. McLagan, D.S.; Monaci, F.; Huang, H.; Lei, Y.D.; Mitchell, C.P.J.; Wania, F. Characterization and Quantification of Atmospheric Mercury Sources Using Passive Air Samplers. *J. Geophys. Res. Atmos.* **2019**, *124*, 2351–2362. [CrossRef]
86. Jeon, B.; Cizdziel, J.V. Can the MerPAS Passive Air Sampler Discriminate Landscape, Seasonal, and Elevation Effects on Atmospheric Mercury? A Feasibility Study in Mississippi, USA. *Atmosphere* **2019**, *10*, 617. [CrossRef]

© 2020 by the authors. Licensee MDPI, Basel, Switzerland. This article is an open access article distributed under the terms and conditions of the Creative Commons Attribution (CC BY) license (http://creativecommons.org/licenses/by/4.0/).

Article

Mercury Challenges in Mexico: Regulatory, Trade and Environmental Impacts

Bruce Gavin Marshall [1,*], Arlette Andrea Camacho [2], Gabriel Jimenez [1] and Marcello Mariz Veiga [1]

1. Norman B. Keevil Institute of Mining Engineering, University of British Columbia, 517-6350 Stores Road, Vancouver, BC V6T 1Z4, Canada; gjimenez15@gmail.com (G.J.); veiga@mining.ubc.ca (M.M.V.)
2. Coordinación para la Innovación y Aplicación de la Ciencia y la Tecnología-Facultad de Medicina, Universidad Autónoma de San Luis Potosí, Avenida Sierra Leona No. 550, Colonia Lomas Segunda Sección, 78210 San Luis Potosí, Mexico; arlette.camacho@hotmail.com
* Correspondence: bruce.marshall@ubc.ca

Abstract: Primary artisanal mercury (Hg) mining in Mexico continues to proliferate unabated, while official Hg exports have declined in recent years amid speculation of a rising black market trade. In this paper, an assessment of primary Hg mining in Mexico was conducted, with a focus on four sites in Querétaro State. Atmospheric Hg concentrations were measured at two of those sites. In addition, trade data was examined, including Hg exports from Mexico and imports by countries that have a large artisanal gold mining (AGM) sector. Results showed that while annual Hg production in Mexico has ramped up in recent years, official Hg exports reduced from 307 tonnes in 2015 to 63 tonnes in 2019. Since 2010, mercury exports to Colombia, Peru and Bolivia have represented 77% of Mexico's total Hg trade. As the large majority of Hg trade with these countries is apparently destined for the AGM sector, which is contrary to Article 3 of the Minamata Convention, there is evidence that increased international scrutiny has led to an increase in unregulated international transfers. Atmospheric Hg concentrations at the mines show dangerously high levels, raising concern over the risk of significant health impacts to miners and other community members.

Keywords: primary artisanal mercury mining; Mexico; Hg trade data; atmospheric mercury concentrations; Minamata Convention

Citation: Marshall, B.G.; Camacho, A.A.; Jimenez, G.; Veiga, M.M. Mercury Challenges in Mexico: Regulatory, Trade and Environmental Impacts. *Atmosphere* **2021**, *12*, 57. https://doi.org/10.3390/atmos12010057

Received: 20 November 2020
Accepted: 31 December 2020
Published: 31 December 2020

Publisher's Note: MDPI stays neutral with regard to jurisdictional claims in published maps and institutional affiliations.

Copyright: © 2020 by the authors. Licensee MDPI, Basel, Switzerland. This article is an open access article distributed under the terms and conditions of the Creative Commons Attribution (CC BY) license (https://creativecommons.org/licenses/by/4.0/).

1. Introduction

On 16 August 2017, the Minamata Convention of Mercury entered into force for the first block of countries (Parties) that had both signed and ratified the international accord. Out of a total of 128 States (Signatories) that have signed the Convention, to date 124 countries have ratified the accord, with Pakistan being the last on 16 December 2020. This agreement, under the umbrella of the United Nations Environment Program (UNEP), provides a regulatory framework with the aim to "protect human health and environment from anthropogenic emissions and releases of mercury (Hg) and mercury compounds." Specifically, there are imperatives to establish strict controls on mercury trade, gradually stop primary mercury mining and limit the use of mercury or mercury compounds in manufacturing processes, including chlor-alkali production and vinyl chloride monomer used to make polyvinyl chloride-PVC [1]. However, it is important to point out that the Minamata Convention only provides guidelines for the parties to follow, with the implementation of regulatory solutions solely on the responsibility of member countries.

One of the first objectives of the Minamata Convention of Mercury was to push for the phase-out the supply of mercury-added products by 2020, a directive that was officially ratified in September 2019 [2]. This includes the manufacture, import and export of batteries, switches, fluorescent lamps, cosmetics, pesticides, barometers, and thermometers, as well as discouraging the use of mercury in dental amalgams. Another important target is limiting the use of mercury by artisanal gold miners (AGM), who number between 16–20 million and operate in more than 70 countries worldwide [3].

Artisanal gold miners are the world's largest users of mercury, who apply rudimentary amalgamation techniques to recover gold, causing severe impacts to human health and the environment [4–6]. In developing countries, artisanal mining activity plays a crucial role in rural or remote areas, where diversification of the local economy is limited or irregular [4,7]. Poverty and a lack of opportunities leads low-income or unemployed rural inhabitants to pursue AGM, where a gram of gold is currently worth approximately US $61. In addition, the implementation of lockdown measures forced by measures to contain the COVID-19 pandemic have left hundreds of thousands of people unemployed, which only exacerbates the situation.

Worldwide, AGM annually produces approximately 400 tonnes of gold or 12% of total global production, generating US $24 billion in revenue [3]. Concomitantly, AGM is the largest source of anthropogenic mercury emissions, with an estimated average loss of 2058 tonnes of mercury being used and annually released into the environment [8], both from fluvial and atmospheric emissions, which accounts for 37% of total Hg emissions worldwide [9]. However, AGM miners continue to use mercury to produce gold despite well-known environmental and health impacts, as it is accessible, relatively cheap and easy to use [10].

For the implementation of the Minamata Convention directives in each country, the importance of reducing and eliminating the supply and trade of mercury is paramount to being able to regulate its use. As of August 2017, the development of new primary mercury mines has been banned and existing mines have 15 years to complete a total phase-out [9]. In addition, there are restrictions for the use of mercury in product manufacturing and disposal, as well as new regulations to control the import and export of mercury between Parties and Non-Parties [9]. According to the U.S. Geological Survey (USGS), in 2019 approximately 4000 metric tonnes of mercury were produced worldwide, with China (3500 tonnes) and Mexico (240 tonnes) being the top countries, while smaller amounts were produced by Tajikistan (100 tonnes), Peru (40 tonnes), Argentina (30 tonnes), Kyrgyzstan (20 tonnes) and Norway (20 tonnes) [11]. While Indonesia is also considered to be one of the world's largest mercury producers and exporters, there is little information on Hg production. However, UN COMTRADE data showed that Hg exports from Indonesia ramped up to 680 tonnes in 2016 from 284 tonnes in 2015 and 0.81 tonnes in 2014, before dropping down to 152 tonnes in 2017, then only 29 tonnes in 2018 and 13 tonnes in 2019. With an estimated 1 million artisanal gold miners working in 27 provinces in Indonesia [12,13], it appears that Indonesian mercury production is principally used for domestic consumption.

Although it is widely known that Mexico is a significant primary mercury producer, the USGS categorizes it as an export country that "reclaims mercury from Spanish colonial silver-mining waste" [11]. Article 3 of the Minamata Convention states that any mercury produced must only be used in the manufacturing of mercury-added products in accordance with Article 4 (e.g., batteries, fluorescent lamps, cosmetics, pesticides, etc. which are to be phased out in 2020, apart from the continuing use of dental amalgams), in manufacturing processes in accordance with Article 5 (e.g., chlor-alkali production, which is to be phased out in 2025) or be disposed of in accordance with Article 11 [1].

While the Minamata Convention is explicit that mercury is not to be exported for its use in artisanal gold mining, an INECC (Mexican Institute of Ecology and Climate Change) report in 2017 examining the challenges, needs and opportunities to apply the Minamata Convention in Mexico, stated that the main destination of mercury produced in Mexico was for exports principally to Latin-American countries that use it in artisanal gold mining [14]. It is apparent that the increase in Mexican mercury production has occurred as a result of export bans imposed by the United States and the European Union since 2011. In 2010, UN COMTRADE data showed that global mercury imports and exports had been 2600 tonnes and 3200 tonnes, respectively. However, by 2015, global Hg imports and exports had decreased to 1200 tonnes and 1300 tonnes, respectively, while mercury production kept on increasing. Since then, in order to avoid scrutiny due to Minamata Convention compliance, there has been increased evidence of informal or illicit transfers, especially involving Indonesia, Colombia and Mexico [15].

Primary mercury mining typically uses artisanal methods, including the roasting of cinnabar (HgS) ores in rudimentary wood-fired ovens, which heats and condenses the released mercury in the form of metallic mercury. Cinnabar is a mercury sulfide mineral composed of 85% mercury and 15% sulfur, which upon calcination releases mercury vapors. Due to the rudimentary method used by artisanal mercury miners, the mercury vapor contaminates the surrounding local environment and also the lungs of workers operating the ovens, who generally do not use any personal protective equipment [16]. It is well known that exposure to mercury vapor enters into the lungs and circulatory system, causing accumulation in the kidneys and brain, leading to serious neuro-cardiovascular problems [4,17–19].

In this paper, we conducted an assessment of primary mercury mining in Mexico, with a focus on four sites in Querétaro State, including Camargo, Bucareli, San Gaspar and Plazuela. This included a brief history of mercury mining in Mexico and a description of the key stages of the mercury mining process, as well as an analysis of the Hg supply and trade situation with several countries in South America that have a large AGM sector, leading to examination of the regulatory control suggested by Minamata Convention directives. In addition, the environmental impacts of primary mercury mining in Mexico are highlighted by atmospheric Hg concentrations measured at mine sites in two different municipalities in Querétaro. There is also a discussion of the economic alternatives that could be promoted in the region to substitute the destructive practices associated with primary mercury mining.

2. Material and Methods

2.1. Field Study and Data Research

Field observations were conducted in the Sierra Gorda region of Querétaro, Mexico, between July 2015 and February 2017 to complement preliminary information obtained in 2014. The first findings indicated that there were four main areas of influence at Sierra Gorda where artisanal mercury miners (AMMs) obtained cinnabar from underground mines and then used rudimentary technologies to produce metallic mercury at primitive processing centers. This information was corroborated by the Secretariat of Sustainable Development (SEDESU) in Querétaro, which confirmed that irregular activities from local miners had been taking place at Sierra Gorda. They indicated that the main activity of these miners was the informal production of primary metallic mercury. Furthermore, SEDESU provided the contact information for some community leaders in order to obtain access and observe its mining and community activities.

In August 2015, a mercury assessment was conducted in the areas of Camargo and Plazuela in Peñamiller Municipality, as well as Bucareli and San Gaspar in the municipality of Pinal de Amoles (Figure 1). This assessment was designed to gather basic information on extraction processes in the underground mines and what techniques were being used at the processing centers. In order to accomplish those objectives, interviews were conducted with owners of mine concessions, AMM leaders and technicians in the processing centers. These interviews were qualitative in nature, which used informal discussions with the participants to better understand the mining and processing methods employed, safety precautions or lack thereof, inherent risks involved, production rates, number of workers involved, legalities and typical salaries for workers.

In March 2016 and February 2017, site investigations were conducted at an artisanal mercury mine located close to Camargo (Figure 2). The site investigations included atmospheric mercury sampling of the area using a Jerome J405 atomic absorption spectrometer (AAS) in 2016 and 2017, as well as a Lumex RA-915M AAS in 2017. While the mercury sampling in 2016 focused on measuring Hg concentrations in and around the cinnabar distillation ovens in the processing area, the 2017 sampling began first in the center of

Camargo village at the police station and then continued on to the mine and mercury processing area (Figure 3). These atmospheric Hg concentrations were then compared to other measurements made in June 2016 at La Soledad mercury mine in Pinal de Amoles Municipality, Querétaro.

Figure 1. Map of study area showing the location of active primary mercury mines in the State of Querétaro, including Camargo and Plazuela in Peñamiller Municipality and Bucareli and San Gaspar in Pinal de Amoles Municipality. The map has been modified using an image sourced from Via Michelin.

Figure 2. Artisanal primary mercury mine in Camargo, Peñamiller Municipality, State of Querétaro Arteaga, Mexico. Photo taken in February 2017 by Bruce Marshall.

Figure 3. Wood-fired ovens at an artisanal primary mercury mine in Camargo, Peñamiller Municipality, State of Querétaro Arteaga, Mexico. Photo taken in February 2017 by Bruce Marshall.

Data research was also conducted looking at mercury exports from Mexico, as well as mercury imports by countries in South America that have a large artisanal gold mining sector.

2.2. Study Area

The town of Camargo is located in Peñamiller Municipality at an elevation of 1755 m and only 8.6 km by road from Peñamiller, which is the capital of the municipality. Approximately 80% of the territory of Peñamiller Municipality lies within the Sierra Gorda Biosphere of Querétaro.

Camargo has a population of 852 inhabitants, whose overall literacy rate is very low (8.1%). As there are few economic activities in the region, unemployment is generally high, with only 44% of the men and 15% of the women employed [20]. INEGI data has shown that both Peñamiller and Pinal de Amoles municipalities have a high poverty index, with approximately 35% of the populations in extreme poverty [20]. The main economic drivers of both municipalities include small shops and businesses, tourism and mining, as well as a small portion of agriculture and livestock (most of it is used for domestic consumption) [20]. The area is very arid and the terrain is desertic with thin layers of soil, which makes extensive agricultural production a significant challenge. Due to the lack of opportunities and low-income jobs, a high percentage of young people has been migrating to the U.S. in search of a better life.

Although the municipalities of Peñamiller and Pinal de Amoles have some rich deposits of various minerals, including hydrothermal veins of gold, silver, lead, zinc, copper and antimony [21], mercury has always been the most exploited. In 2017, the Instituto Nacional de Ecología y Cambio Climático reported that the State of Querétaro had a total of 19 working mercury mines, producing approximately 102 tonnes/a. The mine in Camargo in Peñamiller Municipality produced one-fifth of that or ~20 tonnes/a, while San Gaspar in Pinal de Amoles Municipality produced approximately 16 tonnes/a [22]. However, in 2019, a new report stated that Querétaro State had 189 mines registered with the Secretariat of Economy, which apparently produced a total of 804.6 tonnes of mercury in 2017 [23]. Furthermore, only four of those mines corresponding to two mining concessions had valid permits with SEMARNAT (Secretariat of the Environment and Natural Resources), which is the authority in charge of carrying out the environmental impact assessment and authorizing the mining activities for the exploitation, exploration and benefit of mercury. Although these permits were set to expire in November and December of 2020, they had been issued prior to the Minamata Convention coming into force [23]. It is important to remember that the development of new mercury mines have been banned for all countries that ratified the Convention and that existing mines have 15 years to close their operations [9].

There is concern that on-going production of mercury in the region will compromise nature conservation areas, especially mines in close proximity to the Sierra Gorda Ecological Reserve, which was declared a Biosphere Reserve by UNESCO in 2001. Sierra Gorda is centered in the northern third of the Mexican state of Querétaro and extends into the neighboring states of Guanajuato, Hidalgo and San Luis Potosí. Within Querétaro, the Sierra Gorda ecosystem extends from the center of the state starting in parts of San Joaquín and Cadereyta de Montes municipalities and covers all of the municipalities of Peñamiller Pinal de Amoles, Jalpan de Serra, Landa de Matamoros and Arroyo Seco, for a total of 250 km^2 of territory [24]. The reserve constitutes one of the most diverse natural areas in Mexico and is home to a number of threatened wildlife species and 15 types of vegetation. In addition, it has been estimated that the Sierra Gorda region has among the largest cinnabar reserves in the world, even considering other old and active mercury mines such as Almadén (Spain), Nuevo Almadén (California, EU), Hidrija (Slovenia), Huancavelica Register (Peru), Virginia Quindien (Colombia) and the Province of Kweichow (China).

2.3. The Mercury Mining Process

Informal mercury mining in Mexico consists of excavating underground shafts in mountainous areas that have rich volcanic deposits in search for cinnabar (mercury sulfide) ores. Crews of artisanal mercury miners (AMM) work hundreds of meters inside the mountains following cinnabar deposits in tunnels 0.5 to 2 m wide. The miners typically use explosives purchased on the black market to blast mineralized quartz veins, which they then extract using pneumatic hammers or, in some cases, simple picks and shovels. The work is extremely physically demanding and often the youngest men are the ones who work inside the mines. After blasting and extraction, the material is then put into sacks and brought to the surface, while the waste rock is used to fill old tunnels.

Once outside of the mine, the cinnabar ore is transported to the processing center where a crew of AMMs manually break up the rock with hammers to reduce the pieces to 1.5 to 4 cm in diameter. The visual selection process separates the rocks with some red (cinnabar) stains from the waste material (an experienced mine supervisor further screens the waste pile to recover valuable material). All of the waste rock is then piled into heaps in open areas nearby. The selected material is then crushed manually or in a jaw crusher and the classification process is done using a 3–5 mm steel screen.

The next step involves roasting the cinnabar ore in old-fashioned retort furnaces which are composed of bricks, cement and steel cylinders to receive the condensed mercury. In the roasting process, heat is applied to the cinnabar sulfide ore, raising the temperature above its mercury boiling point (356.7 °C), which creates an oxidized reaction (O_2) to form sulfur dioxide (SO_2):

$$HgS + O_2 \rightarrow Hg + SO_2 \qquad (1)$$

Artisanal miners also mix quicklime (CaO) with the cinnabar ore to improve the mercury vapor formation according to the equation:

$$4HgS + 4CaO \rightarrow 4\,Hg + 3CaS + CaSO_4 \qquad (2)$$

This allows the mercury to be liberated in vapor form together with water and other substances of the process. Then, all the vapors pass to a cooling system, where mercury is the first element to be condensed into liquid and the other vapors are either captured or vented into the surrounding air. The mercury is then collected and filtered to remove impurities (dark film and scum), leaving 99% pure mercury ready to be packaged. This mercury has adequate enough quality to be directly used in the artisanal gold-silver mining sector; however, for its application in many products such as thermometers, electric switches, barometers, etc., it needs to be further re-distilled.

2.4. History of Mercury Mining in Mexico

For centuries, Mexico has mined cinnabar ores, culminating in the production, trade and use of mercury, beginning in 1555 [25]. In 1557, the Spanish merchant Bartolome de Medina introduced the Patio Process of Mercury Amalgamation in Mexico, which solved the problem of recovering low-grade silver ores. Silver mines quickly adopted this process and silver production boomed in Mexico [25,26].

In 1821, when Mexico declared its independence from the Spanish monarchy, the silver industry became one of the key economic drivers of the new government, which required a steady supply of mercury. In order to meet these requirements, the government eliminated old, established mercury supply restrictions that had been implemented by Spain and promoted the domestic mining of cinnabar [25]. Authorities offered benefits such as tax elimination and rewards when mines annually produced more than 90 tonnes of mercury. As a result of these incentives, old mines re-initiated operations and new mercury mines were gradually opened in the Mexican territory [25].

During the Post-Colonial period (1821–1920), domestic mercury production and mercury imports were used to meet the supply demands for amalgamating silver [25]. However, at the beginning of the twentieth century, cyanide processes for leaching gold and silver were introduced in Mexico [25]. This new technology increased the recovery of both gold and silver, maximized production, eliminated the use of mercury and decreased the environmental impacts caused by mercury use and release [27].

By 1921, Mexican silver mines gradually adopted cyanidation and substituted the Patio Process. Nevertheless, local and international markets required mercury for other applications. According to Castro-Diaz [25], primary mercury production between 1922 and 1967 totaled 18,000 tonnes, with an annual average of 400 tonnes. In 1968, the Mexican Commission of Mining Development (Comision de Fomento Minero) reported a total of 1119 mercury mines operating throughout the country. The Mexican States that had the highest concentration of mines were: Querétaro: 322 mines; Durango: 214 mines; Zacatecas: 212; and San Luis Potosí with 100 mines [25].

From 1968 to 1994, primary mercury production decreased to an average of less than 300 tonnes/a, totaling 7700 tonnes for the 26-year period, with Querétaro being the largest producer [25]. Between 1990 and 1992, the Mineral Resources Council of Mexico (CRM) reported 83 mines operating in the country, including in Querétaro [28], Durango, EdoMex, Guanajuato [27], Guerrero, San Luis Potosí and Zacatecas [29]. Then, in 1994, as large-scale or medium-scale mining companies started exploring more lucrative gold, silver or copper deposits, primary mercury mines did not report any production for that year. Between 1994 and 2013, the mining authorities did not report any primary mercury production whatsoever, with no official registration of any informal or small-scale activity [25].

In 2010, the Mexican Secretariat of Economy reported the existence of 314 metallic mercury mines in the country, although most of them were either inactive or abandoned [25]. The states with the most mercury mines were Querétaro with 75 sites and San Luis Potosí with 56 sites [25]. However, in recent years there have been reports about the potential reactivation of the primary mercury industry in Mexico [22,23,30].

Since 2010, Mexico has been increasing its role as one of the largest suppliers of mercury in the world, together with China and Indonesia. Prior to this, the USA and Spain were the largest mercury exporters over a span of decades, until the European Union banned exports in 2010 and the United States quickly followed suit. However, up until recently, official communication by the Mexican government has contradicted this information. The Secretariat of Economy and the mining government office have repeatedly stated that Mexico has not produced metallic mercury from primary mines since 1995, only formally recognizing production obtained through secondary processes, which has been occurring in Mexico for more than a hundred years [25]. Tailings accumulated from the Patio Process during the Colonial and Post-Colonial eras have presented profitable opportunities for those companies capable of reprocessing waste materials for mercury production, mainly in the State of Zacatecas. However, the estimated annual production of mercury generated from the reprocessing of waste materials (20–24 tonnes) does not match Mexico's total annual production of 240 tonnes [11], which in itself could be vastly underestimated [23], and also does not align with recent annual exports of 230–300 tonnes/a (UN Comtrade). It appears that a significant portion of Mexico's mercury exports come from primary mercury mining; the vast majority of which is unreported and unregulated [15,23].

3. Results

3.1. Assessment of the Primary Mercury Mining Process in Mexico

The results from our assessment showed that the mining and mineral processing methods used in the underground mines at Camargo, San Gaspar, Bucareli and Plazuela were very rudimentary. It was apparent that all of these mines have planning and mine design issues, due to a lack of health and safety standards applicable to the mining metallurgical industry. Typical problems faced by AMMs include mine access, roof control, poor ventilation and illumination, rock slides or cave-ins, rudimentary material transportation, lack of personal protective equipment, as well as general health and safety issues.

For example, it was found that miners at Camargo and San Gaspar used simple winches and railway wagons to transport the cinnabar ore out of the shafts, while Bucareli miners transported the ore material in wheelbarrows, which was then loaded onto their backs.

In terms of processing, the centers at Camargo, San Gaspar, Bucareli and Plazuela all use rudimentary old-fashioned ceramic retort furnaces to roast the cinnabar ore and extract metallic mercury. The AMMs prefer this kind of furnace, due to its simple method of extraction and low capital cost. According to the miners, this type of furnace should address specific heavy-duty requirements in order to be efficient. It must be simple to construct, relatively unaffected by thermal extremes during the roasting process and strong enough to withstand the mechanical stresses of loading and unloading of the ore material.

There is also evidence that some miners process the cinnabar ore with small retort furnaces in the backyards of their homes. All mercury obtained during the week by each informal miner is packaged in empty 600 mL plastic soda bottle containers, which weigh 8 kg each. These soda bottle containers are the vessels in which metallic mercury has been traded for years.

In regards to legalities of the mercury concession holders in Peñamiller and Pinal de Amoles municipalities, it was verified that they are all national Mexicans, possess mining titles that were granted before the Sierra Gorda was declared a Biosphere Reserve, and have their mining taxes paid up to date. In other words, they are all meeting the requirements imposed by the mining regulations in Mexico. However, the concessions also have further obligations required by Mexican mining law which include exploration, mine safety and environmental protection mitigation conducted in accordance with mining metallurgical industry standards, as well as allowing for periodic inspections of their operations by the Secretariat of Energy and Mines and the Secretariat of the Environment and Natural Resources. The concession holders who were interviewed stated that "they didn't have the resources and the technical knowledge to follow all of the legal requirements" and that "the rules should not be the same for small operations as for medium and large-scale mining companies".

Interviews with the AMM leaders revealed that there are more than 1000 miners directly involved in this activity in Camargo, San Gaspar, Bucareli and Plazuela, with more than 40,000 people involved indirectly. In addition, the leaders explained that concession holders have been sub-leasing the mines to local organized artisanal miner groups, albeit with varying agreements in each area. For example, in exchange for full control of the mining operation, concession holders have negotiated a royalty of between 15% and 20% of the final weekly or monthly mercury production. In other cases, groups of miners agreed to pay a percentage of the final production, as well as allowing preference for the concession holder to buy a part or all of the final product. In other cases, the concession holder becomes one of the members of the group of miners and the production is divided equally among the members. However, it is important to note that these organized groups of miners do not constitute any type of legal entity, mining company or mining association. In other words, this informal economy is not recognized, regulated or protected by the State. The structure of these organized groups of miners can range from a simple and functional small mine to a larger and more sophisticated operation.

As stated by the artisanal mercury miner leaders, product value has a strong correlation not only with demand, but also with the market price in Mexico as established by the middlemen (this price is not the same as the international mercury price) and the production rate per week. Overall, the average income for AMMs has fluctuated between $2000 and $3000 Mexican pesos (US $105 to US $158) per week before expenses.

3.2. Mexican Hg Supply and Trade

In 2013, it was estimated that Mexico had reserves of approximately 56 thousand tonnes of mercury primarily in the form of probable Hg ore reserves, secondary mercury from mining wastes, and in the chlor-alkali industry, with lesser amounts as by-product mercury from the base metals production sector and secondary production from recycling [25].

Primary mercury mines contribute nearly 75% of these total reserves or 42,000 tonnes, while secondary mercury from mining wastes contributes approximately 25% or 14,000 tonnes [25]. A 2017 report by the government National Institute of Ecology and Climate Change (INECC) indicated that eight of the 31 Mexican states have mercury mines that feed the national trade in dental amalgams, lamps and raw materials for artisanal gold mining, as well as the growing export [14].

UN COMTRADE (United Nations International Trade Statistics Database) data show that Mexico exported a record of approximately 1994 tonnes of metallic mercury from 2010–2018, while also importing 64 tonnes from a variety of sources [31]. The reported exports rose sharply from approximately 26 tonnes/a in 2010 to a peak of 307 tonnes/a in 2015, before declining to 267 tonnes/a in 2016, 200 tonnes/a in 2017 and 230 tonnes/a in 2018 (Figure 4). Although UN COMTRADE data does not show any figures for Mexican Hg exports in 2019, SIAVI (Tariff Information System via Internet) data from Mexico showed an export total of approximately 63 tonnes/a in 2019 [32] (Figure 4). However, it is likely that the vast majority of the estimated production total of 240 tonnes reported by the United States Geological Survey [11] was destined for export, underlying a critical discrepancy with the trade flow data.

The large majority of the mercury exported between 2010 and 2018 (1668 tonnes or 83% of the total) was sent to countries in South America, while the rest was divided between Central America and the Caribbean (6%), Asia (5%), Africa (3%) and North America and Europe with less than 1% each [31].

In an interview with a Mexican mercury exporter in 2017, he declared that there were only a few exporters in the whole country at the time and he was earning US $32,000/month with his business, exporting mercury principally to Colombia and some other Latin American countries.

Since 2010, 42 countries have bought mercury from Mexico. However, mercury exports to Colombia, Peru and Bolivia have represented 77% of the total trade (1545 tonnes of metallic mercury) (Figure 4). During the period 2010–2018, Colombia bought a total of 456 tonnes of mercury from Mexico (22% of the total), with the highest totals being in 2014 (116 tonnes out of a total of 127 tonnes that Colombia imported from all countries) and 2015 (115 tonnes out of a total of 133 tonnes). In that 9-year period, 58% of Colombia's total mercury imports came from Mexico. However, there is no data regarding imports of mercury by Colombia for 2019 [31].

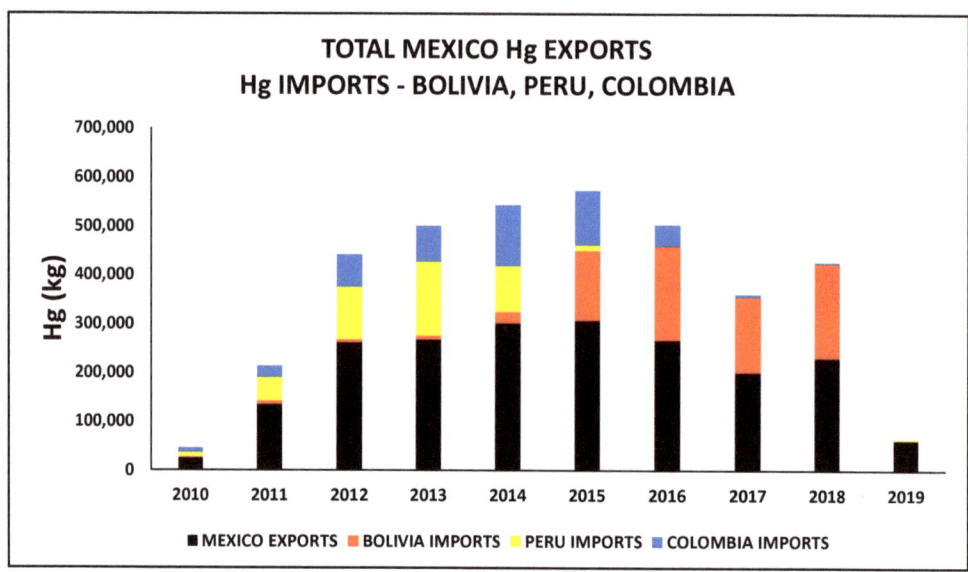

Figure 4. UN COMTRADE data showing the total amounts of mercury exported by Mexico globally for the years 2010–2019 and the import totals of Mexican mercury by Bolivia, Peru and Colombia during the same time period.

From 2010–2014, the second largest receiver of Mexico's mercury exports was Peru with 415 tonnes (27% of total Mexican Hg exports and nearly 60% of all Peruvian Hg imports), whereby 107 tonnes were imported in 2012 (out of a total of 111 tonnes from all countries), 147 tonnes in 2013 (out of a total of 169 tonnes) and 94 tonnes in 2014 (out of a total of 102 tonnes). Subsequently, Peru's total mercury imports drastically declined to 11 tonnes in 2015, with all of it coming from Mexico, while Peru at the same time exported 2.4 tonnes of mercury to Bolivia. Then, in 2016, due to agreements associated with the Minamata Convention, Peru proudly declared that it had stopped all Hg imports (and exports). However, by 2017 Peru began importing mercury once again, totaling 5.2 tonnes for the year, with 0.2 tonnes coming from Bolivia and 5 tonnes from Japan. Then, in 2018, Peruvian Hg imports only reached a total of 0.034 tonnes, with all of it coming from Mexico. However, in 2019 Peruvian imports showed a total of 7.9 tonnes, the highest since 2015, with 1.9 reported tonnes coming from Mexico and the rest from Japan. In total, 59% of Peru's total Hg imports in the period 2010 to 2019 came from Mexico. It is also important to point out that while Peru imported 7.9 tonnes of mercury in 2019, it also exported 91.5 tonnes, with most of it (approximately 88 tonnes) going to the Netherlands [31].

At the same time that Peru's Hg imports declined significantly in the last few years in comparison to the highs reached from 2010 to 2014, Bolivia dramatically increased its official imports from Mexico over a three-year period, from 12 tonnes in 2014 to 139 tonnes in 2015 (out of a total of 143 tonnes imported from all countries), and then 193 tonnes in 2016 (out of a total of 224 tonnes). In 2017, the import total reduced slightly to 180 tonnes, with most of this coming from Mexico (155 tonnes), while smaller amounts came from Spain (10.9 tonnes), Japan (6.9 tonnes) and India (6.7 tonnes). Then, in 2018, Bolivia imported a total of 196 tonnes, with the majority (194 tonnes) coming from Mexico, although a very small amount was also imported from Peru (0.003 tonnes). Other Hg imports in 2018 came from Turkey (2.6 tonnes), USA (0.005 tonnes) and Spain (0.001 tonnes). In total, over the period 2010 to 2018, 91% of Bolivia's Hg imports came from Mexico. However, there is no data regarding imports of mercury by Bolivia for 2019 [31].

This data shows that while Peru decreased its mercury imports from Mexico from over 150 tonnes/a to zero over a three-year period, Bolivia increased its imports approximately

23 times during the same period. In addition, other Latin-American countries such as Panama, Argentina and Paraguay have also been increasing their Mexican imports since 2015, with Argentina importing 12 tonnes in 2019, its highest total since it imported 15 tonnes in 2015. For the 128 countries that signed the Minamata Convention, 31 have traded metallic mercury with Mexico in the last 7 years (this includes 16 parties and 15 signatories). However, since 2018, it has become increasingly difficult to find accurate import/export data, as a large part of the mercury trade has clearly gone underground to avoid scrutiny.

In addition to a lack of current trade data, it has also been noticed that UN COMTRADE data [31] shows discrepancies between the amount of mercury exported by a country and the amount that was imported. For example, an analysis of the Mexican Hg imports by Bolivia, Peru and Colombia highlighted some interesting results. For Peru, the difference between the amounts that Mexico exported and what was imported by Peru ranged from 0% to 24% for the years 2014 to 2015, while both countries reported 0 exports for 2016 and 2017. However, in 2018, Mexico reported exporting 207 kg to Peru, while Peru reported importing only 34 kg. Then, in 2019, while Peru reported an import total of 1.9 tonnes of mercury from Mexico, Mexico did not report any mercury exports (or imports).

For Colombia, the discrepancies between the imported amounts from Mexico and the export totals reported by Mexico were only 3% to 6% for the period from 2014 to 2016, but then in 2017, 2.5 tonnes was the reported export amount from Mexico, while the imported amount by Colombia was 4.6 tonnes (a difference of 45%). Then, in 2018, both countries reported the exact same amount that was exported by Mexico and imported by Colombia: 2.0 tonnes. However, in 2019, neither country declared any mercury trade whatsoever [31].

Finally, for Bolivia there was a large difference in the reported amounts through UN COMTRADE data in 2014, when Mexico reported a total of approximately 24 tonnes exported to Bolivia, while Bolivia reported a total of approximately 12.1 tonnes imported from Mexico, which is a difference of nearly 50%. During the years 2015 to 2018, when Bolivia imported on average 171 tonnes per year from Mexico, the export/import variances ranged from 3% to 16%, with the latter high occurring in 2018 when Mexico reported an export total of 163.3 tonnes to Bolivia, while Bolivia reported a total of approximately 194 tonnes from Mexico. Then, in 2019, there has been no information whatsoever whether Bolivia imported any of the reported 62.6 tonnes of mercury that Mexico officially exported to the world [32].

3.3. Atmospheric Hg Concentrations at Primary Mercury Mines in Mexico

Atmospheric mercury concentrations measured at primary mercury mines in Camargo in Peñamiller Municipality and La Soledad in Pinal de Amoles Municipality showed extremely high levels that workers are exposed to over long periods (Table 1). At the Camargo Mine, in close proximity (0.5–6 m) to the retort furnaces, measurements taken in March 2016 using a Jerome J405 showed atmospheric mercury concentrations that ranged from a low of 9360 ng/m^3 to a high of 62,940 ng/m^3, the latter measured next to the mercury condenser while the furnace (oven) was in operation. It is important to point out that the ovens at the Camargo mine are located outside in a well-ventilated area, which may reduce the Hg concentration levels measured by the spectrometer.

Table 1. Mercury atmospheric concentrations (ng/m^3) in the proximity of two primary mercury mines in Querétaro State, Mexico.

Mercury Mine	Date	Mercury Analyzer	Site Location	Measurements (n)	Min-Max Atm. Hg (ng/m^3)	Avg. Atm. Hg (ng/m^3)
Camargo, Peñamiller Municipality	March 2016	JEROME J405	Workers' changeroom	1	3360	-
			Inside ovens 2 days after processing	2	13,650–14,830	14,240
			3–6 m from ovens operating	6	9380–22,190	18,738
			Next to mercury condenser while oven in operation	2	55,110–62,940	59,025
			At the Camargo police station	1	29	-
		LUMEX RA 915M	En route to the mine	4	132–200	168
			1 km from the mine	1	307	-
			10 m from the processing area	6	1800–8000	5050
	February 2017		Entrance to the mine where the cinnabar is extracted	1	6000	-
			20 m from the mine	1	2290	
			10 m from the mine	1	2780	
			At the entrance to the mine	1	2850	
		JEROME J405	3–10 m from the ovens	8	1800–15,000	7037
			1 m from the retort furnace, where cinnabar ore had been roasted 24–36 h before	2	37,460–50,000	43,730
			Ore feeding door	1	22,020	
			Next to mercury condenser	2	45,821–51,760	48,790
La Soledad, Pinal de Amoles Municipality	June 2016	JEROME J405	5–10 m away from mercury ovens in operation	4	3600–47,100	18,475
			1–4 m away from ovens in operation	8	131,900–438,700	237,412

In comparison, measurements taken within 1–10 m of the retort furnaces at the same Camargo mine in February 2017 using a Jerome J405 showed a low of 1800 ng/m^3 and a high of 51,760 ng/m^3, the latter taken right next to the condensation pipes in front of the ovens where the cinnabar ore had been roasted 24–36 h before and where fresh cinnabar was waiting to be processed (Table 1). It was surprising how similar the Hg concentrations were between the two visits in March 2016 and February 2017, especially considering the latter measurements were taken more than 1 day after roasting a batch of cinnabar ore.

The study in February 2017 also measured atmospheric mercury concentrations using a Lumex 915M beginning at the police station in Camargo, which showed a level of 29 ng/m^3 (Table 1). It is important to note that prior to arriving in Camargo we stopped on the highway from Extoraz, Peñamiller, to get a background Hg concentration, which showed a measurement of 2–5 ng/m^3. Normally, current levels of mercury in outdoor air, except for regional "hot spots," are generally in the order of 2–10 ng/m^3 [33].

From the police station in Camargo, atmospheric Hg concentrations were then monitored using the Lumex while driving the short distance to the mercury mine (approximately 3-4 km), which showed concentrations between 132–300 ng/m^3. At the entrance to the mine tunnel where the cinnabar is extracted, concentrations using the Lumex that averaged 6000 ng/m^3, while the Jerome showed concentrations averaged 2850 ng/m^3 (Table 1).

When approaching the area of the ovens where the cinnabar ore is roasted to produce liquid mercury, only the Jerome was used, as there was concern that the Lumex could become contaminated. At a distance between 3 and 10 m from the ovens, the Hg concentrations using the Jerome were in the range of 1800 to 15,000 ng/m^3. At a distance of approximately 1 m from the ovens, where mercury had been processed 24–36 h prior and fresh cinnabar was waiting to be inserted into the kilns, atmospheric mercury levels spiked to 37,460–50,000 ng/m^3, which is 19 to 25 times above the tolerable concentration of 2000 ng/m^3 for long-term inhalation exposure to elemental mercury vapor [34].

At La Soledad mercury mine in Pinal de Amoles Municipality, where atmospheric mercury concentrations were measured in March 2016 using a Jerome J405 (Table 1), at a distance of 5 to 10 m from ovens in the midst of roasting a batch of cinnabar ore, Hg levels ranged from 3600 to 47,100 ng/m^3, which are comparable to the concentrations measured at the Camargo mine. However, when reaching 1–4 m from the operating ovens, Hg concentrations ranged from a low of 131,900 ng/m^3 to a high of 438,800 ng/m^3, which is 66 to 219 times higher than the long-term inhalation exposure limit.

Figure 5 shows Hg atmospheric concentrations at both Camargo and La Soledad mines in relation to distance from ovens in operation, as well as ovens at Camargo 24–36 h after processing. As the mercury processing area at La Soledad mine is in a closed area with poor ventilation, the Hg concentrations are much higher than at Camargo Mine, which is located in an open area with good air flow. However, at both mines Hg concentrations decrease in relation to increased distance away from the ovens, with similar concentrations being measured at 8–10 m.

At Camargo Mine, it was surprising to see how high the mercury concentrations were 24–36 h after processing, especially within 0.5 m of the ovens. Right next to the mercury condenser, Hg concentrations were 45,821–51,760 ng/m^3, which were very similar to levels measured when the ovens were operating, varying from 55,110 to 62,940 ng/m^3 (Table 1, Figure 5). As many workers are present to collect the mercury from the ovens the day after processing, including cleaning out the old residues and preparing the kilns for further roasting of cinnabar ore, this data shows that the risk of exposure does not dissipate particularly quickly over time, especially in close proximity to the ovens.

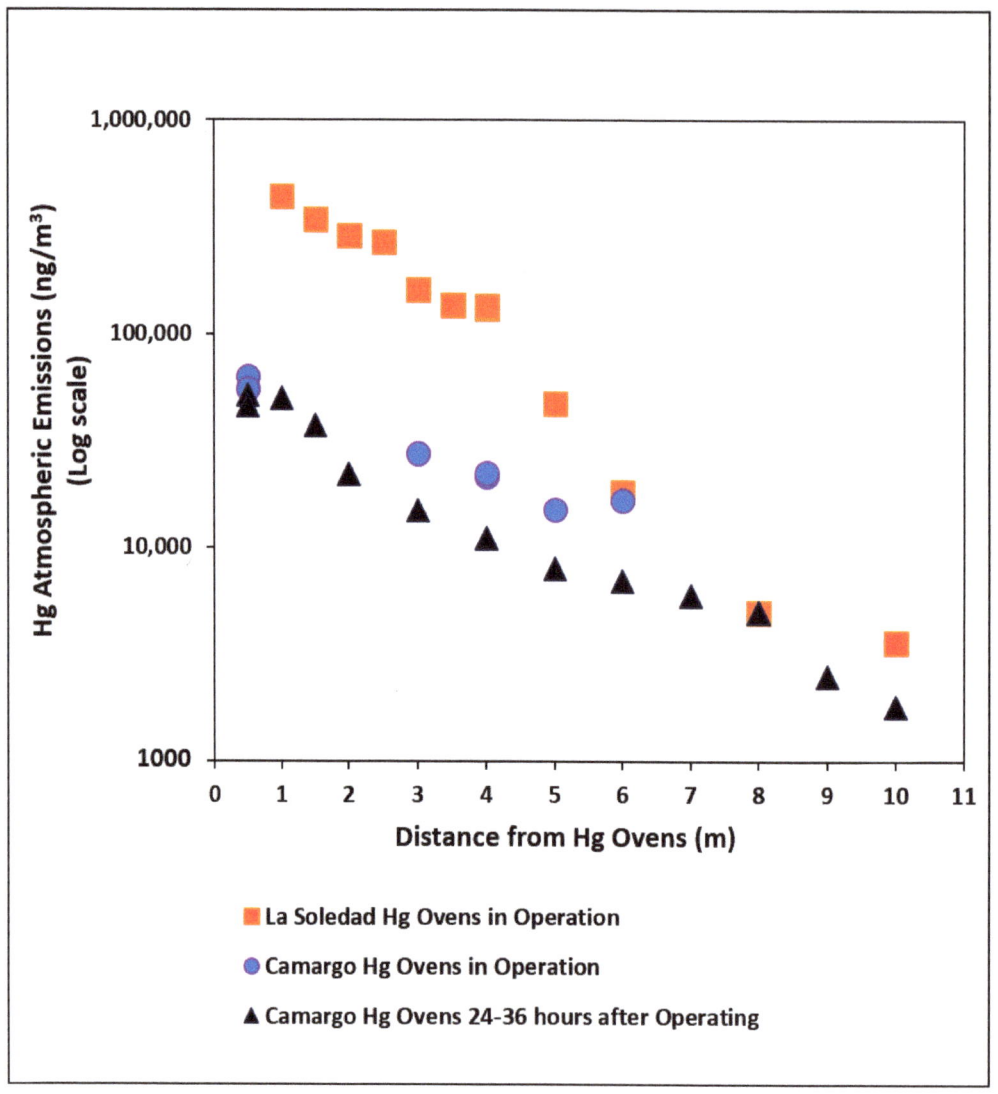

Figure 5. Mercury atmospheric emissions (ng/m^3) at Camargo and La Soledad mines in relation to distance from ovens in operation, as well as ovens at Camargo 24–36 h after processing.

4. Discussion

4.1. Mexican Mercury Supply Chain and International Markets

In 2017, after the Minamata Convention entered into force for signatory parties that signed the agreement four years prior, the global trade in mercury had shifted significantly due to export bans put in place by the United States and the European Union (EU). Not only had the center of the mercury trade shifted from the Northern Hemisphere to Asia and South America, but the main sources of mercury for the trade had moved from chlor-alkali industries to production stemming from artisanal mercury mines.

UNEP's 2017 report on the supply, trade and demand of global mercury revealed that the amount of available chlor-alkali residual mercury on the market, which is used in various industrial processes, had reduced significantly since 2011, largely due to the export bans. In the EU alone, those prohibitions removed an estimated 650 tonnes of mercury off the market [15].

In order to make up for the demand, artisanal mercury mines in Mexico and Indonesia, two of the world's largest producers, began ramping up production starting in 2011. In Mexico, although official primary mercury mining ceased in 1994 due to low global mercury demand and prices, the Mexican Geological Service in 2011 reported that three mercury mines appeared to have reopened and were working intermittently [25]. By 2015, it was estimated that annual mercury production had reached 400–600 tonnes, which would require significant production provided by several hundred artisanal mines [15]. In 2015, 306 tonnes were exported internationally through legal channels, with suspected additional exports made illegally as well.

Studies from the University of Querétaro and the University of San Luis Potosí, a neighboring state, have estimated that 300–400 tonnes of mercury are extracted in Querétaro each year [30]. One visiting group of researchers in 2017 corroborated this estimate, stating that approximately 1000 miners were producing nearly 300 tonnes of mercury per year in Querétaro [35], with a large majority of that production earmarked for export. However, according to the latest report by the Instituto Nacional de Ecología y Cambio Climático [23], it appears that Hg production in Querétaro could be more than 800 tonnes per year, with the vast majority of that production coming from unregulated mines. For Minamata Convention parties like Mexico, which ratified the commitment with the Convention on 29 September 2015, new primary mercury mines were prohibited after the Convention came into force in August 2017, with all existing mines mandated to be phased out within 15 years of that date [9].

As the trade data in this study showed, 83% of the Mexican Hg exports between 2010 and 2018 were destined for countries in South America, with Bolivia, Colombia and Peru comprising 77% of that total. It is also clear that the majority of the imported mercury by these countries ends up in the AGM sector, where it is used to amalgamate gold, generating the world's largest source of mercury pollution. The UNEP report in 2017 estimated that artisanal gold miners globally use the highest amount of mercury (37%), while the second highest use is for the production of vinyl chloride monomer (26%), occurring mainly in China and likely using mercury produced within their own borders [15]. As Mexico, Bolivia, Peru and Colombia have all signed and ratified the Minamata Convention, the trade of mercury for gold amalgamation contravenes Article 3 of the Convention, which states that mercury cannot be used for this purpose [1].

The drastic reduction in Mexican Hg exports in 2019 and the reduced total imports of mercury by countries like Bolivia, Peru and Colombia signal that increased scrutiny regarding the official transfers and illegal use of mercury have obfuscated the transparency of international transactions, thereby driving suspicion of a thriving black-market trade. However, in terms of official databases like UN COMTRADE and SIAVI from Mexico, it is also possible that there are discrepancies with the data due to the following reasons: (a) the statistical manner in which re-exports and transshipments or transiting goods have been treated; (b) errors in transferring information from manual documents to digital; (c) lack of clarity with regard to the actual origin and destination of goods; (d) inaccurate coding of commodities (i.e., mercury compounds instead of listing as mercury); and (e) the occurrence of undocumented shipments, especially goods passing through bonded warehouses or Foreign Trade Zones [15].

It is the occurrence of undocumented shipments that is the most concerning, given that the official global mercury trade has reduced significantly over the past 15 years, while the demand continues to remain strong. For example, in 2005, 3400 tonnes of mercury were traded worldwide, while in 2017 only 984 tonnes were traded globally, while demand was estimated to be between 4500 and 5000 tonnes [36]. In addition to the reasons for data discrepancies mentioned above, including the uncertain quality of data reported to the UN COMTRADE and data transparency issues, there is also evidence of a 2-tier mercury price caused by the export bans, which encourages illegal transfers. This includes corruption schemes involving the re-selling of waste mercury from chlor-alkali plants as pure mercury destined for AGM operations around the world (i.e., the DELA affair in Germany) to circumventions around the export ban in China [15,36].

The major shift in international mercury trading hubs over the past 9–10 years, combined with increased scrutiny over the source and destination of mercury shipments, has fueled a significant increase in informal and illicit transfers among countries, including Indonesia, many African nations, Mexico, Colombia and Bolivia, among many others. In a country like Mexico, these illicit transfers from the producer (AMM) to the consumer (e.g., AGM in Bolivia, Peru or Colombia) require the facilitation of middlemen, who are able to somehow circumnavigate regulatory controls to expedite export shipments. However, a consequence of all of this is an increase in mercury price, which has risen nearly 9-fold in the last 10 years. In 2010, the price in Mexico was US $17/kg, while in 2020 it has skyrocketed to more than US $186/kg [37]. In countries such as Colombia, Peru, Ecuador and Brazil, the price in AGM areas can reach US $350/kg [10].

4.2. Mexican Artisanal Mercury Miners and Their Future

A combination of high unemployment and a lack of economic opportunities in Querétaro municipalities, such as Peñamiller and Pinal de Amoles, have led to the proliferation of poorly regulated and illicit primary mercury mine operations in the Sierra Gorda region that pump out hundreds of tonnes of mercury annually. Although there are 189 concessions in Querétaro registered with the Secretariat of Economy [23], the actual number of mines could be much higher. Information culled from our interviews with AMMs in Camargo, Plazuela, Bucareli and San Gaspar from 2014–2017 showed that during that period it was estimated that there were more than 100 retort furnaces in the region, which was likely an underestimate.

For the artisanal mercury miners, working either in the mines extracting cinnabar ore or manning the ovens to produce metallic mercury offers a better standard of living than other menial labor, as AMMs earn 2 to 5 times more than the standard minimum wage in Mexico. However, as international pressure mounts from the United Nations Environment Programme and Mexican government authorities to clamp down on "new" illicit operations that were not already legal and in existence prior to ratification of the Accord, the livelihood of AMMs becomes increasingly uncertain.

Apart from the miners themselves, it also remains unclear how apparently "legalized" concessions have been extracting cinnabar for the purposes of producing primary mercury without adequate oversight administered by the Mexican mining and environmental authorities. Considering that the vast majority of mercury mines occur in two of the poorest municipalities in the State of Querétaro, with high rates of unemployment, low salaries and few economic opportunities, the lack of supervision and enforcement may have a political basis. Given the economic reality of the region, it is without question that Hg mines and the mercury they produce provide an important source of livelihood for these communities.

According to SEDESU in Querétaro, the increase in mercury price, as well as the increase in gold price for the AGM sector, has caused the surge in reactivation of old mines to extract cinnabar and produce liquid mercury. Although the concessions have been granted to people or groups of people who know the traditional and rudimentary methods, they do not have the financial capacity to improve their practices or provide the necessary mining infrastructure. Nonetheless, many of these sites do not meet the minimum requirements established by the mining authorities. Furthermore, as mercury is a heavy metal and a pollutant, its exploitation must be conducted with security measures, as well as with strict environmental stewardship. As with the AGM sector, mining and environmental authorities seem reluctant to get involved, due to a lack of resources, politics, public protest and an understanding that informal employment is such a strong driving factor in economic sustenance for a vast majority of the population.

Similar to other developing countries, informal employment has been and continues to be a significant part of the Mexican economy. According to the International Labour Organization (ILO) and the National Institute of Statistics and Geography (INEGI), it was estimated that informal employment in Mexico increased to 59% of the total workforce in 2013 [38]. The study found that approximately 30 million people had informal employment in Mexico. The Mexican mining industry is not exempt from this phenomenon, as the ILO report indicated that the rate of informal employment in the mining and quarries industries was approximately 15% in 2013. Considering the economic fallout caused by the coronavirus, where 12 million Mexican workers have lost their jobs in 2020 over the past 9 months, it is clear that informal employment has likely risen sharply, as families struggle to make ends meet [39].

4.3. Health Impacts and Community Issues of Primary Mercury Mining

The extremely high atmospheric mercury concentrations measured at two different mine sites in Querétaro highlight the dangers that artisanal primary mercury miners working either in the mines or in the processing areas producing metallic mercury are exposed to over long periods of time. The health impacts caused by chronic exposure even to moderate atmospheric Hg concentrations can be very serious, leading to damage to the kidneys and brain, resulting in lasting neurological effects. Although the LOAEL (Lowest Observed Adverse Effect Level) for mercury vapor is between 15,000 and 30,000 ng/m^3, with an 8-h time weighted average (TWA) for occupational exposure set at 25,000 ng/m^3, the NOAEL (No Observed Adverse Effect Level), which is a guideline for public exposure of inorganic mercury vapor, has been established at 1000 ng/m^3 as an annual average [33].

In order to determine the health quality index or hazard quotient (HQ) for workers exposed to mercury vapor, the following equation for chronic exposure could be used: HQ = EC/Toxicity Value or Reference Dose (RfD), where EC (exposure concentration) (ng/m^3) = (CA × ET × EF × ED)/AT. Specifically, CA (ng/m^3) = contaminant concentration in air, ET (hours/day) = exposure time, EF (days/year) = exposure frequency, ED (years) = exposure duration, and AT (ED in years × 365 days/year × 24 h/day) = averaging time [40]. For occupational exposure, the RfD could be set at 25,000 ng/m^3.

In this study, exposure to atmospheric Hg concentrations at the Camargo and La Soledad mines increased significantly in relation to closer proximity of the ovens. At 10 m from the ovens, concentrations averaged approximately 2700 ng/m^3, while at 5 m concentrations averaged 23,000 ng/m^3, at 3 m 67,680 ng/m^3, and at 1 m 244,350 ng/m^3. In this example: the ET might be 8 h a day; the EF is calculated as 240 days/year (at 5 days a week); a typical ED might be 30 years; and the AT is therefore equal to 262,800 (30 years × 365 days/year × 24 h/day).

For calculation of the HQs for chronic exposures at 10 m, 5 m, 3 m and 1 m, the results would be: 10 m = 0.02; 5 m = 0.2; 3 m = 0.6; and 1 m = 2.1. When the Hazard Quotient is >1, then the Toxicity Value has been surpassed and the exposed populations could be at risk. In this case, although it may appear that the workers are only at potential risk when exposed to extremely high Hg concentrations (~244,000 ng/m^3 measured at 1 m from the ovens), the calculation of HQs done in this way could be underestimating the risk, as Hg production typically involves batches of ore to be processed, followed by days without high exposure. Therefore, it could be more appropriate to consider the exposures as acute in nature, which would be calculated using EC = CA, then EC/RfD. In this way, the results would be: 10 m = 0.1; 5 m = 0.9; 3 m = 2.7; and 1 m = 9.8, which are significantly higher and likely better denote the actual risk unprotected workers face when exposed to atmospheric mercury concentrations.

In 2016, field investigations by a team of researchers from the University of San Luis Potosí (UASLP), examining the health and environmental impacts of primary mercury mining at different mine sites, found highly elevated urinary mercury concentrations in miners at Camargo [41]. Urine samples from a total of 103 miners working at the site showed that 99% had Hg urine concentrations above 35 µg/g creatinine, which is the risk threshold for men, although the LOAEL for renal toxicity has been established at 4 µg/g creatinine [42]. While the minimum concentration found was 31 µg/g creatinine, the median was 275 µg/g creatinine, while the maximum concentration was 2599 µg/g creatinine, which is 74 times above the risk threshold. When the urine mercury concentration exceeds 100 µg/L, neurological symptoms can develop, and above 800 µg/L can be fatal [43].

Considering that quite often women and young children can also be involved in the mercury mining operations, often helping with menial tasks or breaking the cinnabar ore into smaller pieces, chronic exposure to elevated atmospheric mercury concentrations for these populations can be especially detrimental. A study investigating mercury levels in the urine of artisanal primary mercury miners, women and children living in Plazuela in Peñamiller Municipality, where there are three mercury mines, found that average concentrations were 53, 35 and 22 µg/g creatinine, respectively, with maximum concentrations of 144, 63 and 37 µg/g creatinine, respectively [16]. As the risk threshold for women and children has been established at 20 µg/g creatinine, these numbers are very concerning, especially when you consider that renal toxicity can occur at concentrations above 4 µg/g creatinine [42].

For these vulnerable populations chronically exposed to dangerous levels of atmospheric mercury concentrations, further health assessments should be conducted to investigate clinical damage to the kidneys or central nervous system. In addition, verification of a health issue for both miners and community members require implementation of personal protective equipment protocols to safeguard against on-going occupational risk.

5. Conclusions

As the gold price remains high and artisanal gold mining using mercury amalgamation proliferates around the world, demand for mercury from countries like Mexico and Indonesia will continue unabated, even with implementation of international agreements like the Minamata Convention. While countries that have signed and ratified the Convention press on with the implementation of National Action Plans to reduce Hg use in products and manufacturing processes, while also finding ways to mitigate against legacy sites of mercury-contaminated wastes, the demand for mercury used to produce vinyl chloride monomer (mainly in China) and in the AGM sector (in more than 70 countries around the world) continues to be an on-going problem with no end in sight. As the United Nations Environment Programme (UNEP) stated in their report: "it is evident that global mercury demand will have to be reduced in parallel with supply, or else supplies–formal or informal–will continue to be generated in one manner or another to meet demand" [15].

As official mercury trade through imports and exports have reduced from historic highs in 2015 to negligible levels in 2019, there is strong evidence of increasing volume in illicit and informal transfers, which is attracting more international attention. The heightened scrutiny of Mexico and Indonesia legal exports since the Minamata Convention came into force in 2017 appears to have been the causal factor in the dramatic reduction of official exports and the vast majority of the mercury trade being pushed underground. As Peter Maxson reported in 2020, it has become clear that the mercury trade is increasingly following the gold trade, which is centered in areas like India, the United Arab Emirates and their intermediaries, and the drug trade, in places like Colombia [36].

Due to the close linkages between the mercury trade and the AGM sector, finding a solution to reduce Hg use is difficult, although the alternatives certainly exist. Through education, capital investment and training, it is feasible to introduce mercury-free, clean and efficient gold extraction methods to AGM communities, as well as taking steps to shorten the gold value chain. What this requires though, is a concerted effort by funding agencies, international partners and national government ministries to curb illegal mercury production, illicit trade and Hg demand through the implementation of better economic and technological alternatives, as well as vigilant enforcement of criminal activities by racketeers.

6. Recommendations and Future Steps

For the artisanal mercury miners in Querétaro, Mexico, a number of incentives are desperately needed to tackle the problems associated with informal and unregulated Hg production using rudimentary methods, as well as the rampant unreported trading of primary mercury. It is clear that task forces need to be set up to ensure that clandestine mines are closed, as well as providing crucial oversight of mercury mines that were operating legally prior to Mexico ratifying the Convention, which also requires enforcement of legislation regarding proper mercury waste management practices.

This includes the implementation of programs to ensure that workers exposed to dangerous levels of atmospheric mercury have access to personal protective equipment, as well as sufficient training and education to learn about new mining and processing technologies. In addition, more health monitoring of both miners and community members exposed to atmospheric Hg concentrations needs to be conducted, including investigation into potential medical issues suffered by newborn children.

On an economic front, before shutting down any mercury mines and putting vulnerable populations out of work, other alternatives need to be introduced, including new industries that create jobs in the region. As Rodríguez-Galeotti argued [44], conventional small or medium-scale mining of available minerals such as gold, silver, lead, antimony and zinc could be viably exploited, although it would still require adequate oversight and regulatory controls to avoid any negative environmental impacts to the Sierra Gorda Biosphere.

Another economic possibility is the idea of developing eco-tourism in the Biosphere, which constitutes the greatest ecosystem diversity found in Mexico. There are 1700 plant species, 30% of all of the butterfly species in the country and 600 vertebrate species, including black bears, macaws, spider monkeys, jaguars, mountain lion, bobcat, margay, ocelot and jaguarundi. However, in order to satisfy the demands and comfort level of tourists, the local infrastructure would have to improved, which would require significant investment by the State.

Although agriculture is currently practiced in the region, mainly for local consumption, selective crops with high export value like oregano and alfalfa could be exploited in a selective way, as well as expanding grape production for an incipient wine industry in Querétaro. However, as Hg concentrations in residential soils close to primary mercury mines have been found to be up to 150 times above the Mexican guideline [16], care would have to be taken to develop commercial agriculture in areas free from contamination, including verification of heavy metal concentrations in the soils and sediments. Tree-farming has also been raised as another possibility, but the poor soils of the desertic region pose a limiting factor for wide-spread success.

Since primary mercury mining has been practiced in the region for hundreds of years, there is also the challenge of social acceptance by locals to move away from this activity, to modify the current rudimentary practices, or to mine another kind of mineral from known deposits in the region. In addition, as mercury miners can make up to 5 times the minimum wage, other economic activities need to be equally as lucrative to entice community endorsement. Shutting down the mercury trade not only affects the miners and their communities, but also the regional economic spin-offs that middlemen and the export supply chain support indirectly.

In order to curb unreported and illegal transfers of Mexican mercury to fuel its use in the AGM sector in countries like Bolivia, Peru and Colombia, it is necessary to understand the dynamics of the illicit trade, assemble inter-agency working groups, develop public-private partnerships, enhance the scale-up of anticorruption task forces, establish better cooperation with law enforcement and support non-governmental organizations on the ground.

In October 2018, the INECC in Mexico submitted a proposal for GEF funding titled "Reducing global environmental risks through the monitoring and development of alternative livelihood for the primary mercury mining sector in Mexico." The concept was approved in December 2018 for GEF funding totaling US $7,035,000 and co-funding of an additional US $51,068,844, although the project is still currently waiting to be implemented. However, once it finally gets underway, this kind of initiative shows excellent promise in being able to provide better oversight of the primary artisanal Hg sector in Mexico, as well as develop other economic alternatives for vulnerable populations.

Author Contributions: This research article was developed in collaboration with all of the authors. Conceptualization: M.M.V., B.G.M. and G.J.; methodology: M.M.V., B.G.M. and G.J.; validation: M.M.V., B.G.M.; formal analysis: M.M.V., B.G.M. and G.J.; investigation: M.M.V., B.G.M., G.J. and A.A.C.; resources: M.M.V.; data curation: M.M.V., B.G.M., G.J. and A.A.C.; writing—original draft preparation: B.G.M., G.J. and M.M.V.; writing—review & editing: B.G.M., M.M.V.; visualization: M.M.V., B.G.M. and G.J.; supervision: M.M.V.; project administration: M.M.V.; funding acquisition: M.M.V. All authors have read and agreed to the published version of the manuscript.

Funding: This research received no external funding.

Institutional Review Board Statement: Not applicable.

Informed Consent Statement: Not applicable.

Data Availability Statement: Not applicable.

Acknowledgments: The authors would like to thank Fernando Díaz-Barriga of the Universidad Autónoma de San Luis Potosí, Beatriz Verduzco Cuéllar from the Universidad Autónoma de Querétaro, and Miriam Loera for their help with support in the field. In addition, we are grateful to Arturo Gavilán García, Miguel Ángel Martínez Cordero and Tania Ramírez Muñoz of the Instituto Nacional de Ecología y Cambio Climático (SEMARNAT) and Alfonso Conde Asiain of the Servicio Geológico Mexicano for personal communications useful to this study. Financial support for this study was provided by the Canadian International Resources and Development Institute (CIRDI).

Conflicts of Interest: The authors declare no conflict of interest.

References

1. UNEP—United Nations Environment Programme. *Minamata Convention on Mercury: Text and Annexes*; UNEP: Nairobi, Kenya, 2013.
2. UNEP—United Nations Environment Programme. Reducing Mercury in Artisanal and Small-Scale Gold Mining (ASGM). 2019. Available online: https://web.unep.org/globalmercurypartnership/our-work/reducing-mercury-artisanal-and-small-scale-gold-mining-asgm (accessed on 25 September 2019).
3. Seccatore, J.; Veiga, M.M.; Origliasso, C.; Marin, T.; Tomi, G.D. An estimation of the artisanal small-scale production of gold in the world. *Sci. Total Environ.* **2014**, *496*, 662–667. [CrossRef] [PubMed]
4. Veiga, M.M.; Baker, R. *Protocols for Environmental and Health Assessment of Mercury Released by Artisanal and Small-Scale Gold Miners*; GEF/UNDP/UNIDO Global Mercury Project: Vienna, Austria, 2004; p. 289.
5. Betancourt, O.; Narváez, A.; Roulet, M. Small-scale gold mining in the Puyango River basin, southern Ecuador: A study of environmental impacts and human exposures. *EcoHealth* **2005**, *2*, 323–332. [CrossRef]
6. Marshall, B.G.; Veiga, M.M.; Kaplan, R.J.; Adler Miserendino, R.; Schudel, G.; Bergquist, B.A.; Guimarães, J.R.D.; Gonzalez-Mueller, C. Evidence of transboundary mercury and other pollutants in the Puyango-Tumbes River basin, Ecuador-Peru. *Environ. Sci. Proc. Impacts* **2018**, *20*, 632–641. [CrossRef] [PubMed]
7. Verbrugge, B. The economic logic of persistent informality: Artisanal and small-scale mining in the Southern Philippines. *Dev. Chang.* **2015**, *46*, 1023–1046. [CrossRef]
8. AMAP/UN Environment. Chapter 3: Methodology for Estimating Mercury Emissions. In *Technical Background Report for the Global Mercury Assessment 2018*; Arctic Monitoring and Assessment Programme: Oslo, Norway; UN Environment Programme, Chemicals and Health Branch: Geneva, Switzerland, 2019.
9. UN Environment. *Global Mercury Assessment 2018*; UN Environment Programme, Chemicals and Health Branch: Geneva, Switzerland, 2019.
10. Veiga, M.M.; Marshall, B.G. Teaching artisanal miners about mercury pollution using songs. *Extr. Ind. Soc.* **2017**, *4*, 842–845. [CrossRef]
11. U.S. Geological Survey—USGS. *Mineral Commodity Summaries 2020*; U.S. Geological Survey: Reston, VA, USA, 2020; p. 200.
12. Balifokus. IPEN Mercury Treaty Enabling Project, Indonesia. 2015. Available online: https://16edd8c0-c66a-4b78-9ac3-e25b63f72d0f.filesusr.com/ugd/13eb5b_e95e7961efb04b6e9248f446629036f2.pdf (accessed on 20 November 2020).
13. Balifokus. Mercury Trade and Supply in Indonesia. 2017. Available online: https://16edd8c0-c66a-4b78-9ac3-e25b63f72d0f.filesusr.com/ugd/13eb5b_bf0b2658eccf40cc9dbbb3a6514e9d64.pdf (accessed on 20 November 2020).

4. Martínez Arroyo, A.; Páramo Figueroa, V.H.; Gavilán García, A.; Martínez Cordero, M.A.; Ramirez Muñoz, T. *Asesoría para la Identificación de Desafíos, Necesidades y Oportunidades para Aplicar el Convenio de Minamata en México*; Instituto Nacional de Ecología y Cambio Climático: Mexico City, Mexico, 2017; p. 114.
5. UN Environment. *Global Mercury Supply, Trade and Demand*; United Nations Environment Programme, Chemicals and Health Branch: Geneva, Switzerland, 2017.
6. Camacho, A.; Van Brussel, E.; Carrizales, L.; Flores-Ramírez, R.; Verduzco, B.; Ruvalcaba-Aranda Huerta, S.; Leon, M.; Díaz-Barriga, F. Mercury mining in Mexico: I. community engagement to improve health outcomes from artisanal mining. *Ann. Glob. Health* **2016**, *82*, 149–155. [CrossRef] [PubMed]
7. Lim, H.E.; Shim, J.J.; Lee, S.Y.; Lee, S.H.; Kang, S.X.; Jo, J.Y.; In, K.H.; Kim, H.G.; Yoo, S.H.; Kang, K.H. Mercury inhalation poisoning and acute lung injury. *Korean J. Intern. Med.* **1998**, *13*, 127–130. [CrossRef] [PubMed]
8. World Health Organization. Exposure to mercury. In *Public Health and Environment*; World Health Organization: Geneva, Switzerland, 2007.
9. Park, J.-D.; Zheng, W. Human exposure and health effects of inorganic and elemental mercury. *J. Prev. Med. Health* **2012**, *45*, 344–352. [CrossRef] [PubMed]
10. National Institute of Statistics and Geography. *Economic Census Data, Mexico in Figures*; National Institute of Statistics and Geography: Mexico City, Mexico, 2019.
11. Servicio Geológico Mexicano. *Anuario Estadístico de la Minería Mexicana*; Servicio Geológico Mexicano: Mexico City, Mexico, 2018.
12. Martínez Arroyo, A.; Páramo Figueroa, V.H.; Gavilán García, A.; Martínez Cordero, M.A.; Ramirez Muñoz, T. *Generar Información Cualitativa y Cuantitativa de las Fuentes Minero-Metalúrgicas en México*; Instituto Nacional de Ecología y Cambio Climático (INECC): Mexico City, Mexico, 2017; p. 217.
13. Instituto Nacional de Ecologia y Cambio Climático (INECC). Anexo de la decisión MC-1/8. Formato de presentación de informes del Convenio de Minamata sobre el Mercurio. In *Información Sobre las Medidas que se han de Adoptar para Aplicar las Disposiciones del Convenio, la Eficacia de Esas Medidas y las Dificultades con que se ha Tropezado*; Instituto Nacional de Ecología y Cambio Climático (INECC): Mexico City, Mexico, 2019. [CrossRef]
14. Velasco Mireles, M. El mundo de la Sierra Gorda. *Arqueol. Mex.* **2006**, *13*, 28–37.
15. Castro Díaz, J. *An Assessment of Primary and Secondary Mercury Supplies in Mexico*; Commission for Environmental Cooperation: Montreal, QC, Canada, 2013.
16. González-Reyna, G. *Riqueza Minería y Yacimientos Minerales de México*; Monografias Industriales del Banco de México, S.A.: Ciudad de México, Mexico, 1947.
17. Camargo, J.A. Contribution of Spanish–American silver mines (1570–1820) to the present high mercury concentrations in the global environment: A review. *Chemosphere* **2002**, *48*, 51–57. [CrossRef]
18. Consejo de Recursos Minerales. Monografía geológico-minera del estado de querétaro. In *Serie Monografías Geológico Mineras (in Spanish)*; Secretaría de Energía, Minas e Industria Paraestatal: Ciudad de Mexico, Mexico, 1992.
19. Consejo de Recursos Minerales. Monografía geológico-minera del estado de zacatecas. In *Serie Monografías Geológico Mineras*; Secretaría de Energía, Minas e Industria Paraestatal, Consejo de Recursos Minerales: Ciudad de México, Mexico, 1991. (In Spanish)
20. UAQ-UASLP. *Plan de Desarrollo Humano y Tecnológico, Sostenible, Para la Zona Produtora de Mercurio del Estado de Querétaro*; Universidad Autónoma de Querétaro, Universidad Autónoma de San Luis Potosí: Queretaro Santiago, Mexico, 2016.
21. UN Comtrade. United Nations Statistical Division. 2020. Available online: http://comtrade.un.org/ (accessed on 15 August 2020).
22. Sistema de Información Arancelaria Via Internet. Available online: http://www.economia-snci.gob.mx/ (accessed on 28 August 2020).
23. World Health Organization. *Chapter 6.9 Mercury, Air Quality Guidelines*; WHO Regional Office for Europe: Copenhagen, Denmark, 2000.
24. International Programme on Chemical Safety. *Concise International Chemical Assessment Document 50: Elemental Mercury and Inorganic Mercury Compounds: Human Health Aspects*; World Health Organization: Geneva, Switzerland, 2003.
25. Spiegel, S.J.; Agrawal, S.; Mikha, D.; Vitamerry, K.; Le Billon, P.; Veiga, M.; Konolius, K.; Paul, B. Phasing out mercury? Ecological economics and indonesia's small-scale gold mining sector. *Ecol. Econ.* **2018**, *144*, 1–11. [CrossRef]
26. Maxson, P. Mercury Material Flow: Supply, Demand and Trade. Available online: http://www.mercuryconvention.org/News/fromtheConvention/MOFlowTrade/tabid/8570/language/en-US/Default.aspx (accessed on 29 September 2020).
27. Mercado Libre Mexico. Available online: https://www.mercadolibre.com.mx/ (accessed on 10 November 2020).
28. International Labour Organization. Informal employment in Mexico: Current situation, policies and challenges. Notes on formalization. In *Programme for the Promotion of Formalization in Latin America and the Caribbean (FORLAC)*; International Labour Organization: Geneva, Switzerland, 2014.
29. Reuters. *Mexico Loses 12 Million Jobs, Workers in Informal Sector Grow*. Available online: https://nationalpost.com/pmn/health-pmn/mexico-loses-12-million-jobs-workers-in-informal-sector-grow (accessed on 10 November 2020).
30. Environmental Protection Agency. *Risk Assessment Guidance for Superfund Volume I: Human Health Evaluation Manual (Part F, Supplemental Guidance for Inhalation Risk Assessment)*; Office of Superfund Remediation and Technology Innovation, EPA: Washington, DC, USA, 2009.

41. Camacho, A.; Rebolloso, C.; Carrizales, L.; Van Brussel, E.; Flores, R.; Díaz-Barriga, F. *Minería Artesanal de Mercúrio—Riesgos en Salud*; University of San Luis Potosí (UASLP): San Luis Potosi, Mexico, 2016.
42. US Centers for Disease Control and Prevention (CDC). *Third National Report on Human Exposure to Environmental Chemicals*; CDC: Atlanta, GA, USA, 2005.
43. Goldman, L.R.; Shannon, M.W. Technical report: Mercury in the environment: Implications for pediatricians. *Pediatrics* **2001**, *108*, 197–205. [CrossRef] [PubMed]
44. Rodríguez-Galeotti, E. La minería de mercurio en México. Boletín de Mineralogía. In *Quadrum Metals & Minerals*; Universidad Autónoma del Estado de Hidalgo, Academia de Ciencias de la Tierra: Pachuca, Mexico, 2006; Volume 17, pp. 29–36.

MDPI
St. Alban-Anlage 66
4052 Basel
Switzerland
Tel. +41 61 683 77 34
Fax +41 61 302 89 18
www.mdpi.com

Atmosphere Editorial Office
E-mail: atmosphere@mdpi.com
www.mdpi.com/journal/atmosphere

www.ingramcontent.com/pod-product-compliance
Lightning Source LLC
LaVergne TN
LVHW070602100526
838202LV00012B/541